Printed in the United States of America

First Printing, 2018

ISBN: 9781796861785

www.PipetteKids.com

Creator
Charlotte Anthony

Editor
Ramaswamy Sharma, Ph.D.

Writers
Charlotte Anthony, Armando Murillo and Alison Doyungan Clark, Ph.D.

Layout & Design
Chase Fordtran

Illustrators
Rebecca Osborne & Angela Gao

Medical Illustrator
Sue Simon

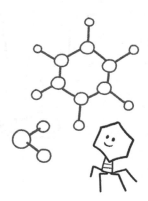

If you need help, be sure to ask an adult or scientist.
You can also email us at help@pipettekids.com

PIPETTE KIDS
MEETING SCIENTISTS OF THE PAST, PRESENT, AND FUTURE

This book belongs to: _____

Draw yourself as a scientist

This is Poly. She wants to be a biomedical engineer so she can invent cool things to help people.

This is Taq. He wants to be a cancer biologist so that he can do experiments that will help cure his grandma's cancer.

Poly and Taq are best friends. They go to the same school.

Their favorite class is science. One day, their teacher Ms. Salinas asks them to write a report on their favorite scientist or inventor.

Ms. Salinas asks Poly and Taq to stay after class to show them a way to find a scientist to write about for their report — a cool virtual reality headset that will let them meet scientists of the past.

Taq and Poly look around the room and point to the microscope. Ms. Salinas explains that although Romans have been looking through glass since the first century, Anton Van Leeuwenhoek is known as the inventor of the microscope.

Anton Van Leeuwenhoek owned a store that sold fabric and he wanted to see which thread had a better quality. At the time, there were no lenses that were capable of doing this. So, he made his own lenses by placing the middle of a small rod of glass in a hot flame creating glass spheres–– which he used as lenses. With his new lenses, he could now see things that were too tiny to see with just our human eyes.

ACTIVITY #1:
MAKE YOUR OWN MICROSCOPE

MATERIALS
- Plastic Cup
- Plastic Wrap
- Rubber Band
- Scissors
- Water
- Leaves or some other specimen of your choice

STEPS
1. Cut a hole in the bottom of your plastic cup using scissors (with the help of an adult).
2. Stretch a piece of plastic wrap over the top of the cup and secure it with a rubber band.
3. Place it through the hole in the bottom of the cup.
4. Pour a little water on top of the wrap.
5. Look at your specimen through the water. The water now acts as a lens.

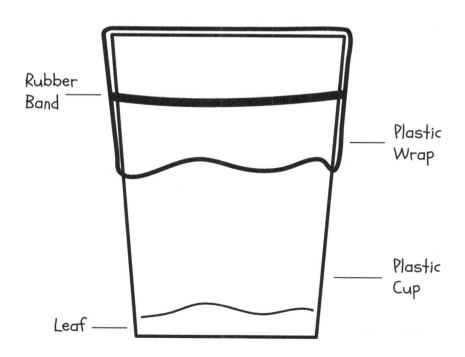

DRAW WHAT YOU SEE

CONNECT-THE-DOTS:
MICROSCOPE

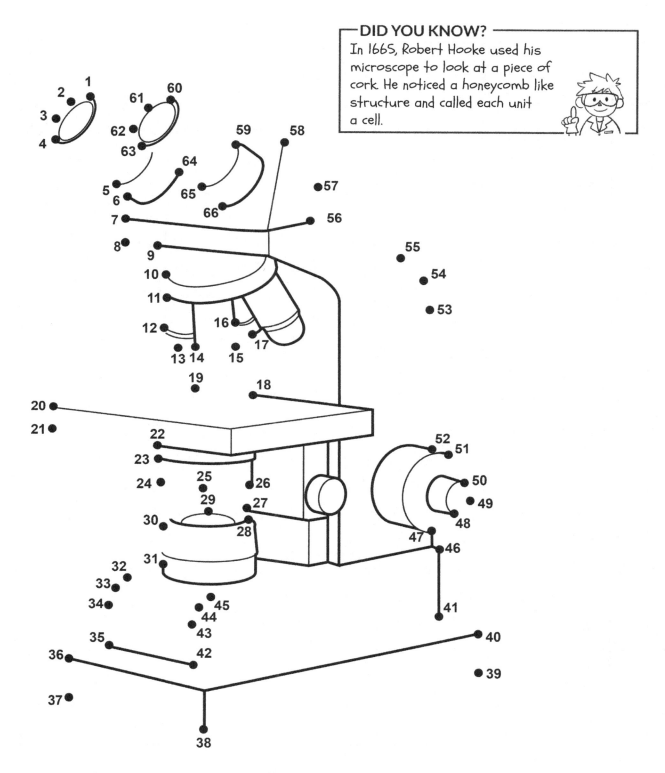

DID YOU KNOW?
In 1665, Robert Hooke used his microscope to look at a piece of cork. He noticed a honeycomb like structure and called each unit a cell.

ANSWER KEY IS ON PAGE 120

Scientists use microscopes to see objects that are too small to be seen by our eyes. Look at the images below and guess what they show. The answer key is on page 120.

1. _____

2. _____

3. _____

4. _____

Anton Van Leeuwenhoek used his microscope to see many things, including bacteria in his teeth.

|

COLOR THE BACTERIA

When scientists look at bacteria under a microscope, we see five major shapes.

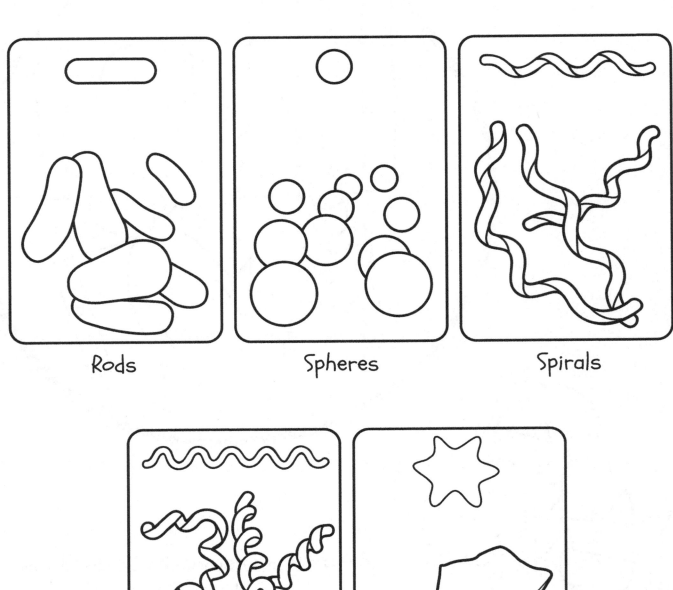

Rods

Spheres

Spirals

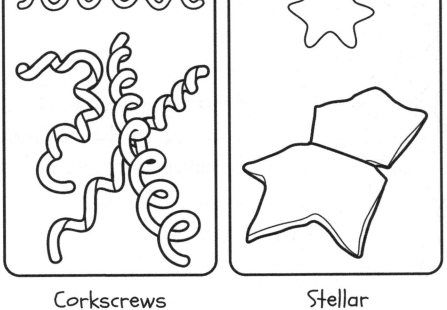

Corkscrews

Stellar

COLOR THE BACTERIA

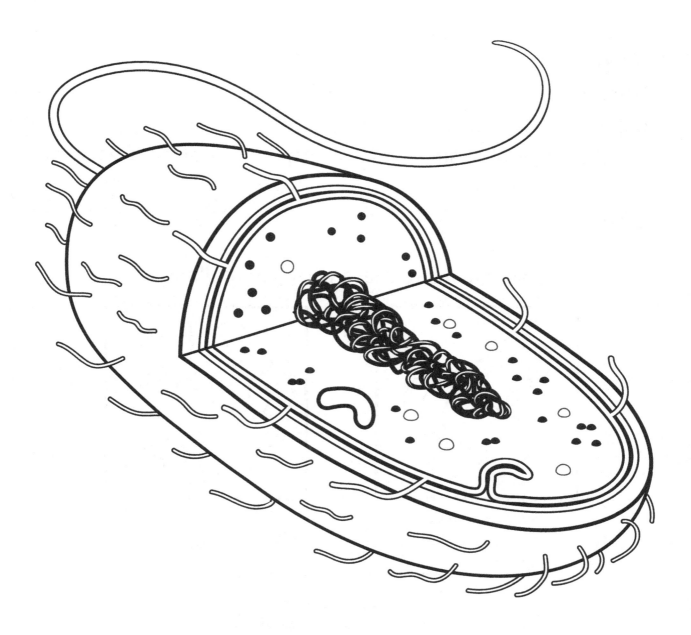

Which shape is this? _____

(Hint: see previous page)

ACTIVITY #2:
Make Your Own Bacteria Balloons

MATERIALS
- Balloons of different shapes and sizes
- Balloon pump (if needed)
- Lengths of wool or small strips of paper (to represent flagella)
- Double sticky tape
- Strings

Can you make these with your balloons?

Streptococcus
pneumoniae

Staphylococcus
aureus

Salmonella typhi

|

WHAT ARE GERMS?

Germs are tiny organisms that can cause us to get sick. They come in a variety of shapes and sizes and can live everywhere from the air to the soil to the water and even in our body. Even though we come into contact with germs every day, our immune system works hard to protect us.

Bacteria are one-cell creatures that get nutrients from their environment in order to live. Ear infections, sore throats, tuberculosis, cavities, and pneumonia are some examples of bacteria-caused illnesses.

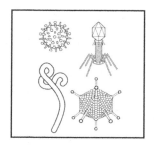

Viruses can't survive very long if they're not inside a living thing like a plant, animal, or person. Whatever a virus lives in is called its host. When viruses get inside people's bodies, they can spread and make people sick. Chickenpox, measles, influenza, and the common cold are some examples of virus-caused illnesses.

Fungi are plant-like organisms that like to live in damp, warm places. They get their nutrition from plants, people, and animals. Athlete's foot, yeast infections, ringworm are examples of fungi-caused illnesses.

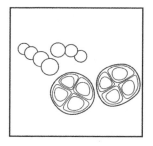

Protozoa are one-cell organisms. They depend on others for food. Protozoa are found virtually everywhere in moist environments. As a group, protozoa are extremely adaptable. African sleeping sickness, malaria and toxoplasmosis are examples of protozoan-caused illnesses.

ACTIVITY #3:
Does soap really make a difference?

MATERIALS

— Glo Germ™ powder or gel

— UV flashlight

— Nearby sink and soap

STEPS

1. Rub Glo Germ™ powder or gel onto your hands, making sure to apply it to all areas like the front, back, and around your fingers.

2. Rinse your hands in cold water and then observe the results in a dark room with a UV flashlight. Rinse for about one minute.

3. What did you see?

4. Now wash your hands with soap and water until all the powder or gel is gone.

5. Reapply the gel and rinse your hands in warm water. What was the difference?

6. Look at your hands under the UV flashlight with the lights off. Do they look cleaner? Or not?

—DID YOU KNOW?—

Hungarian doctor Ignaz Semmelweis wanted to find out why so many women in maternity wards were dying of puerperal fever. He found out that doctors who had recently examined patients in the morgue went straight to the maternity ward to deliver babies had not washed their hands. Semmelweis hypothesized that if the doctors used chlorine to disinfect their hands and instruments, less people would get sick. True enough, hand washing did help and saved countless lives.

Circle which activities Poly and Taq are doing to prevent illnesses.

Sharing drinks

Washing hands

Staying home when not feeling well

Blowing nose into tissue

But not all germs are bad! Poly is getting hungry. She goes to her lunchbox to get her sandwich but is shocked to find mold. Ms. Salinas tells her that mold saves millions of lives each year. The kids look confused. Ms. Salinas tells them to put on their virtual reality goggles to meet Alexander Fleming.

When **Alexander Fleming** came back from vacation, he noticed that a bit of mold was growing into his petri dish. He was researching staphylococcus, a group of spherical bacteria which can cause a variety of diseases or illnesses like food poisoning. Fleming was surprised to find that the mold helped to kill bacteria. He tested his theory and found that mold could be helpful to treat many infections like scarlet fever, pneumonia, meningitis and more.

DID YOU KNOW?

In the past, moldy bread was used to prevent infection by pressing bread against the wound. Honey was also used to dress wounds which we now know contains hydrogen peroxide.

Just like Alexander Fleming's discovery, some other great discoveries have happened by chance.

Archimedes, the Greek mathematician had figured out clever solutions on how to measure things which is why King Hiero decided to ask Archimedes for help on a problem he was having. King Hiero wanted to find out if the crown he had purchased was made of pure gold. Archimedes was stumped. After taking a bath, he noticed that the volume of the water that overflowed was exactly equal to the bulk of the part of the body that he placed in the water. He realized that he could do the same for the gold crown. He ran out of the bathtub shouting Eureka!

Joseph Priestly helped us learn about the element oxygen after he discovered how to make soda water. He learned that there was a gas (which we now know as carbon dioxide) that was heavier than ordinary air.

FUN TIP:
Next time you make a discovery, shout Eureka!

TOOLS OF A SCIENTIST

Scientists use many tools to answer basic questions about how things work and why things are the way they are. Help us match the word with the object by drawing a line to connect them. The answer key is on page 120.

Microscope

Petri Dish

Eye Dropper

Measuring Cup

MORE TOOLS OF A SCIENTIST

Scientists today have lot of advanced tools to help them in the lab such as a PCR machine which can help amplify segments of DNA. We will learn more about DNA soon.

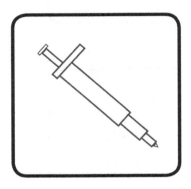

They also use a pipette to help transfer liquids similar to an eye dropper. Pipettes can come in many sizes, such as a micropipette which can help transfer liquids less than 1 ml.

The Erlenmeyer flask was created by Emil Erlenmeyer, a German chemist, in the 1860s. He developed a flask that has a flat bottom and a tapered neck. It is used for mixing solutions and can also be closed with a stopper in order to store samples.

Scientists also have lots of boxes of gloves in the lab. This is to help them avoid contamination and protect their hands from infection. Gloves create a barrier between germs and our hands. They help us keep our hands clean. Nitrile gloves also help protect against some chemical and infectious agents.

SO MANY TOOLS
CROSSWORD PUZZLE

WORD BANK

> GEL BOXES
> CENTRIFUGE
> ERLENMEYER FLASK
> GLOVES
> MICROSCOPE SLIDES
> PETRI DISH
> BEAKER
> PIPETTE
> TEST TUBE RACK
> BUNSEN BURNER
> FUNNEL
> INCUBATOR
> GOGGLES

ACROSS

4. A shallow dish used to culture cells such as bacteria.
5. Glassware with a flat bottom with a tapered neck.
6. A tool used to hold several test tubes at the same time.
7. A pipe with a wide conical mouth and a narrow stem.
9. Device used to grow and maintain microbiological or cell cultures.
10. Container that can accommodate gel trays.
12. Device used to measure weight.
14. Machine used typically to separate fluids of different densities.

DOWN

1. A piece of laboratory equipment that produces a single open gas flame.
2. Used to transfer small amounts of liquids.
3. Thin flat piece of glass used to hold objects for examination under a microscope.
8. Item used to protect eyes.
11. Item used to protect hands.
13. Glassware with a wide mouth and usually a lip for pouring.

ANSWER KEY IS
ON PAGE 120

HOW MANY SCIENCE/SCIENTIFIC TOOLS DO YOU COUNT?

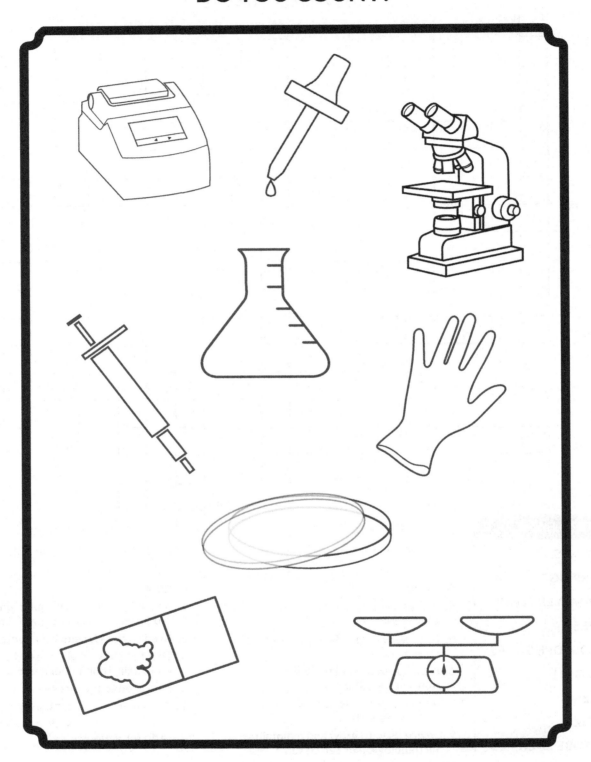

YOUR ANSWER: _____

ANSWER KEY IS ON PAGE 120

MATCH THE SHAPES TO THE SHADOWS

You did a
GREAT JOB!

ANSWER KEY IS ON PAGE 120

|

WORD SCRAMBLE

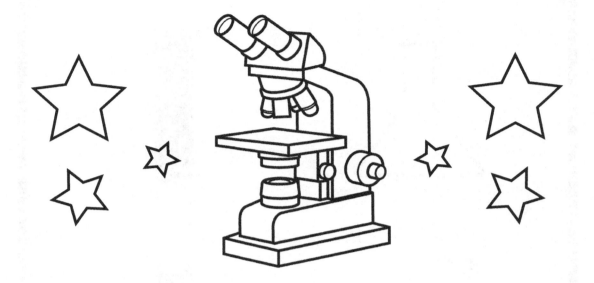

UNSCRAMBLE THE WORDS

GRTCEIFNUE

_____

EIPTPTE

_____

CUOBNATIR

_____

ANSWER KEY IS ON PAGE 120

MEGA MATHLETE

KEY: 🧪 =1 🔬 = 2 ⚗️ =3 🧫 =4

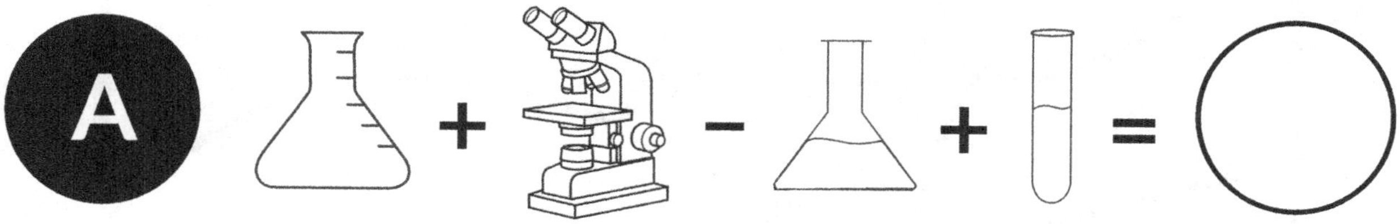

A 🧪 + 🔬 − ⚗️ + 🧫 = ◯

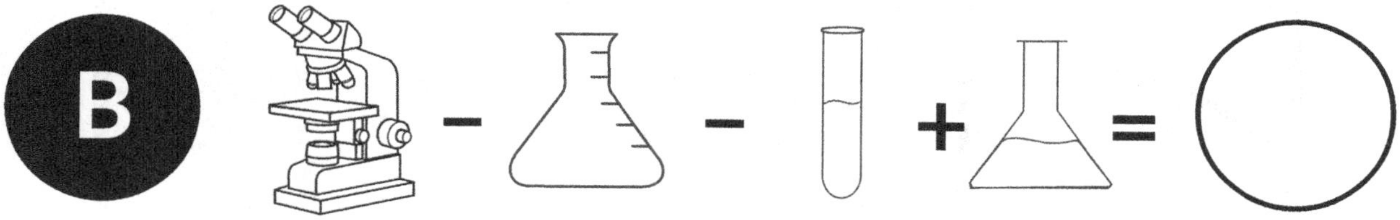

B 🔬 − 🧪 − 🧫 + ⚗️ = ◯

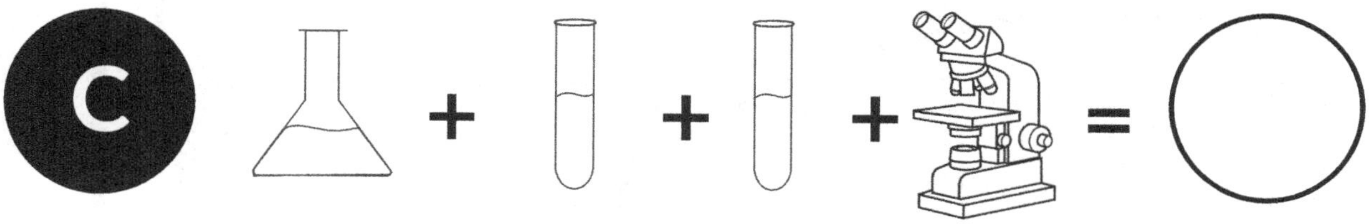

C ⚗️ + 🧫 + 🧫 + 🔬 = ◯

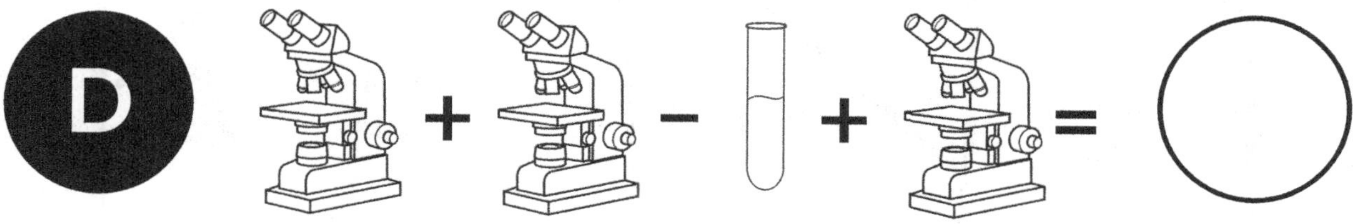

D 🔬 + 🔬 − 🧫 + 🔬 = ◯

ANSWER KEY IS ON PAGE 120

COLOR BY NUMBER PIPETTE

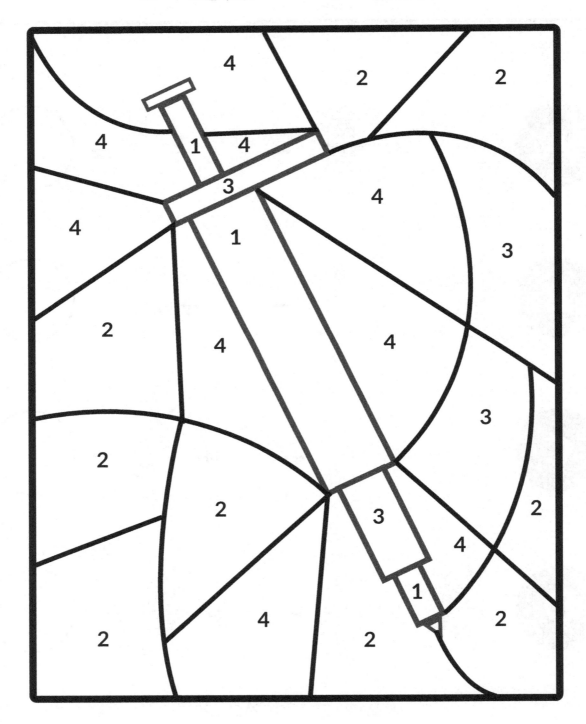

SIX KINGDOMS

There are many different kinds of living things on Earth.
Scientists have grouped them together into six kingdoms.

ANIMALS
(Horses, whales, sponges)

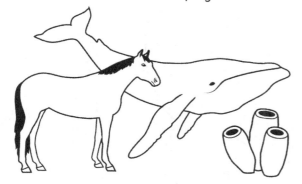

Multicellular, eukaryotic

PLANTS
(Grass, trees, flowers)

Multicellular, eukaryotic

FUNGI
(Mushrooms, molds, lichens, yeasts)

Multicellular and unicellular,
eukaryotic

PROTISTS
(Amoebas, kelp, plankton, paramecia)

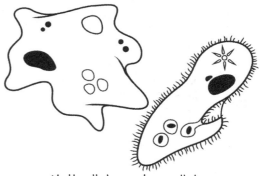

Multicellular and unicellular,
eukaryotic

EUBACTERIA
(Helpful and harmful relationships
with humans)

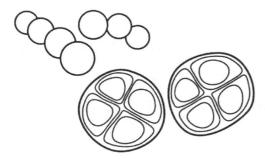

Unicellular, prokaryotic

ARCHAEBACTERIA
(Live in extreme habitats, don't infect
humans)

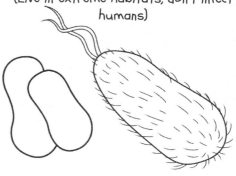

Unicellular, prokaryotic

Bacteria and other living organisms are made up of cells. A cell is the smallest unit of life. Most cells are only visible under a microscope. Scientists like Alison Clark are called "cell biologists." Alison looks at our cells to learn more about the human body.

|

The human body has more than 200 types of cells. These are just a few.

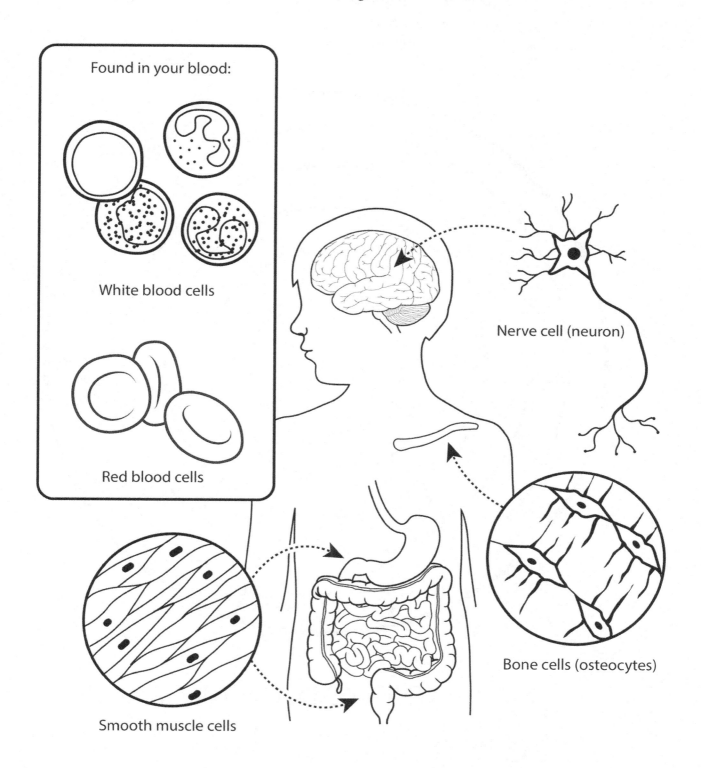

Found in your blood:

White blood cells

Red blood cells

Nerve cell (neuron)

Bone cells (osteocytes)

Smooth muscle cells

WHITE BLOOD CELL MAZE

White blood cells are cells in our immune system that help protect your body. They move throughout our body in our blood and help fight infections by attacking bacteria, viruses, and germs.

DID YOU KNOW?

WHITE BLOOD CELLS ACCOUNT FOR ABOUT 1% OF YOUR BLOOD.

CELL CITY

Animal cells are like miniature cities, containing many different organelles. Color in the cell and its organelles. How many of each organelle can you find? Write the number on the line.

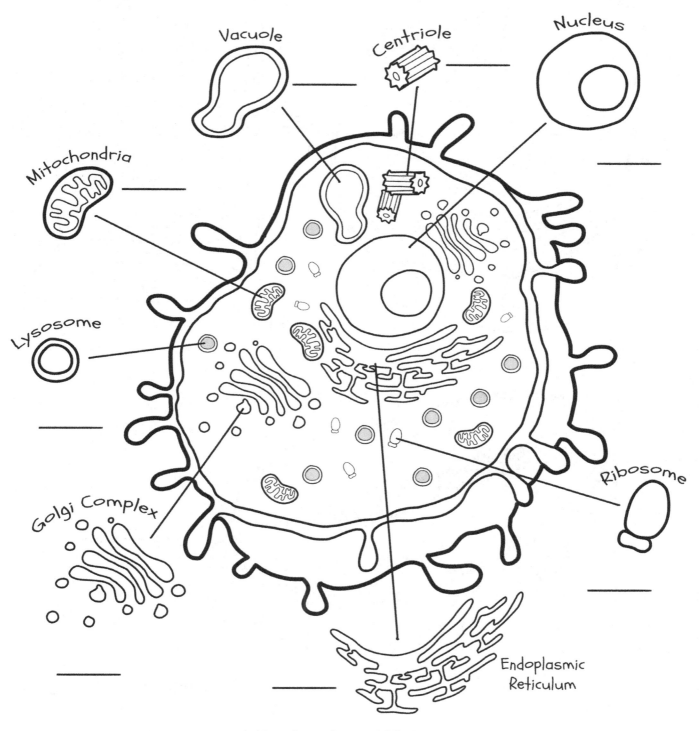

ANSWER KEY IS ON PAGE 120

COLOR THE CELLS

Animal Cell

Plant Cell

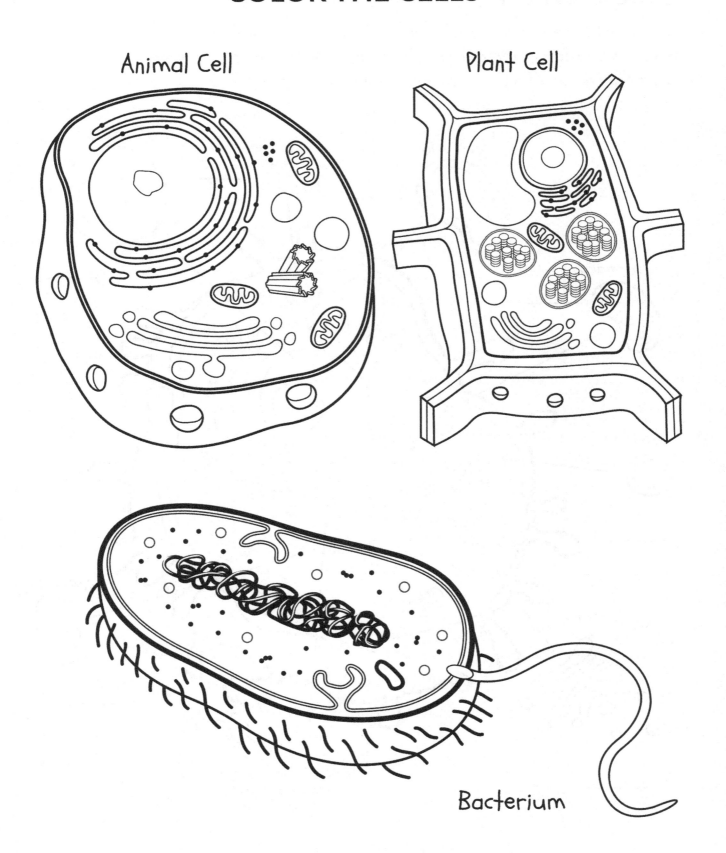

Bacterium

|

Poly wants to know why she has her dad's hair. Taq says it is because of genes.

Genes are a sequence of our DNA (we will learn more about this later) that tell your body to display a trait. Differences in genes are the reasons why humans have so many different eyes, skin, and hair colors.

|

This is Gregor Mendel. He is known as the "father of genetics." He studied how traits can be inherited using pea plants. He saw that when he bred yellow peas and green peas, all the offspring were yellow pea plants. He determined that the yellow trait must be a dominant trait. However, when he bred the new yellow peas with each other, he got both yellow and green pea plants. He realized that green must be a recessive trait. This is important because when we think about ourselves, we learn that we inherit traits from our parents and our grandparents.

DOMINANT VS. RECESSIVE TRAITS

DOMINANT GENE	RECESSIVE GENE
Cleft Chin	No Cleft
Widow's Peak	No Widow's Peak
Dimples	No Dimples
Freckles	No Freckles
Free Earlobe	Attached Earlobe

MOLECULE OF LIFE

1 = BLUE

2 = ORANGE

3 = LIGHT BLUE

4 = GREEN

5 = YELLOW

6 = PINK

7 = RED

8 = PURPLE

Follow the color key to reveal the molecule of life.

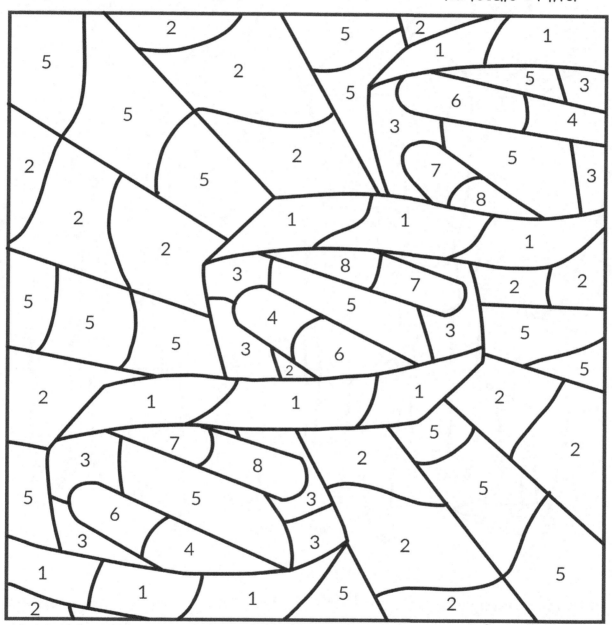

DID YOU KNOW?

DNA stands for "deoxyribonucleic acid." Try saying that 5 times fast!

We didn't always know about DNA. In fact, DNA is one of the greatest scientific discoveries. It all happened at King's College London when Rosalind Franklin obtained images of DNA using X-ray crystallography, an idea first put forth by Maurice Wilkins. Franklin's images allowed James Watson and Francis Crick to create their famous two-strand, or double-helix, model.

|

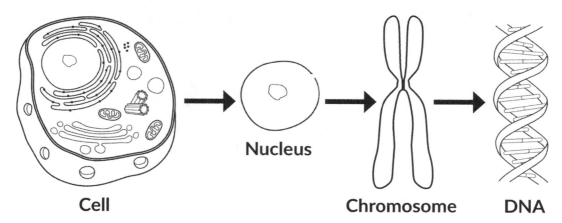

Cell   Nucleus   Chromosome   DNA

Deoxyribonucleic Acid or DNA for short is a long thin molecule that tells cells what to do. The way that our body uses DNA is similar to a computer program. Our cells "read" the genetic code which is made up of these four molecules: A-adenine, T-thymine, C-cytosine and G-guanine. However, there are rules you should know about:

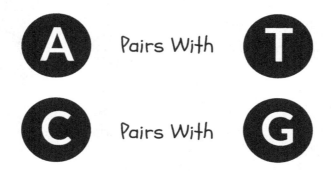

A Pairs With T

C Pairs With G

Even though there are only four different letters, DNA molecules are thousands of letters long. This allows for billions of different combinations.

Here is an example:

CCC CGG TAT ATA TCG CAG GGT CGT
GGG GCC ATA TAT AGC GTC CCA GCA

The sequence of our DNA determines our genes, which determine the traits we have.

DID YOU KNOW?
Even though humans have 3 billion base pairs, only a tiny amount are unique to us.

HOW SIMILAR IS OUR DNA?

99.9%

90%

Did you know that we are 99.9% similar to the person sitting next to us, 96% genetically similar to chimps, 90% as similar to the cat and 60% as similar to a banana!

96%

60%

85%

80%

Source: National Human Genome Institute

ACTIVITY #4:
Extract DNA from Strawberries!

MATERIALS

Extraction Buffer – you will need 2 tsp salt, 2 tbsp dish soap, ½ cup of water. Mix all these components in a cup by stirring

- 1 ziplock bag
- 1 strawberry – remove the green parts
- 1 glass cup
- 1 plastic cup
- Measuring spoons
- Rubbing alcohol
- Cheese cloth or coffee filter
- Funnel
- Tweezers

STEPS

1. Take strawberry and place inside ziplock bag. Gently mash strawberry with your fingers for 2 mins.

2. Measure 2 tbsp of extraction buffer and add to the bag. Make sure you remove as much air as possible when closing the bag.

3. Massage the bag for 1 min to make strawberry extract.

4. Place cheesecloth or coffee filter in the funnel. Then place funnel inside plastic cup.

5. Filter strawberry extract by pouring it onto the cheese cloth or coffee filter. Allow extract to drip into cup. You can also squeeze the cheese cloth or filter paper to speed up the process.

6. Take 1 tbsp of filtered strawberry extract and place inside glass cup.

7. Take 1 cup of alcohol and VERY SLOWLY add down the side of the test tube.

8. You will see fine strands appear in the solution. Congratulations – you have successfully extracted DNA from strawberries!

9. Use tweezers to pick up DNA.

ACTIVITY #5:
How to make DNA Candy Helixes

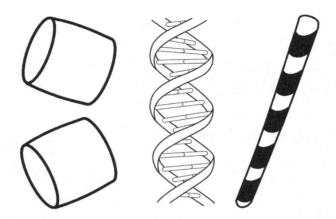

MATERIALS

— Licorice sticks

— Toothpicks

— One bag of marshmallows containing four different colored marshmallows (giant marshmallows are preferable)

STEPS

1. With the help of an adult, sanitize the area you will be working in. Also, wash your hands before working. This is important if you plan on eating your DNA candy helix.

2. Sort out the marshmallows by color. Then assign a color to each of the four DNA bases – adenine, thymine, cytosine, and guanine.

3. Start building your base pairs! Using the toothpick, connect the marshmallows into pairs. Remember the rules of base pairing – adenine must be connected with thymine, and cytosine must connect with guanine. Repeat as many times as you want, but make sure you have multiples of each type of base pair.

4. Connect your base pairs to the backbone. Take the toothpicks with the marshmallows bases and poke the end of the toothpicks through a stick of licorice on each end of the base pair. Repeat this step until you've connected all your base pairs to the licorice backbone. It should look like a ladder.

5. Twist the licorice backbone so your DNA model looks like a spiral staircase.

6. Eat your DNA candy helix if you want to!

ACTIVITY #6:
Build a DNA Sequence Bracelet with your Name

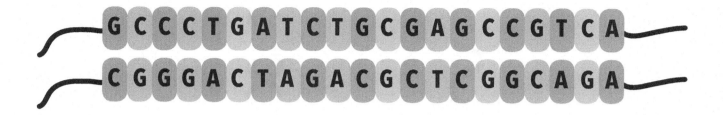

G C C C T G A T C T G C G A G C C G T C A

C G G G A C T A G A C G C T C G G C A G A

MATERIALS

— Plastic beads with four different colors

— Elastic String

STEPS

1. Using the alphabet table, write out your name in DNA code. For example, if your name is Alice, you name in code would be GCC CTG ATC TGC GAG. This will be the sequence of letters you will use to build your bracelet.

2. Sort out the beads by color if they are not already. Then assign a color to each of the four DNA bases – adenine, thymine, cytosine, and guanine.

3. Take two pieces of string and tie them together with a knot. String 1 will be on top, and string 2 will be on the bottom.

4. Start by threading your first bead through string 1. Look at the first letter in your sequence and find the right color of bead to thread.

5. Thread the bead for the matching letter onto string 2. Remember the rules of base pairing – adenine must be connected with thymine, and cytosine must connect with guanine.

6. Repeat steps 4 and 5 until you have finished the sequence for your name.

7. Tie the two strings together to secure the beads. Then tie the ends of the bracelet together.

A - GCC	B - GTC	C - TGC	D - GAC	E - GAG	F - TTC	G - GGC
H - CAC	I - ATC	J - GGG	K - AAG	L - CTG	M - ATG	N - AAC
O - TAA	P - CCC	Q - CAG	R - AGA	S - AGC	T - ACC	U - TAG
V - GTG	W - TGG	X - AAA	Y - TAC	Z - TCT		

Source: www.yourgenome.org

DNA/RNA
What is RNA?

Ribonucleic Acid

Deoxyribonucleic Acid

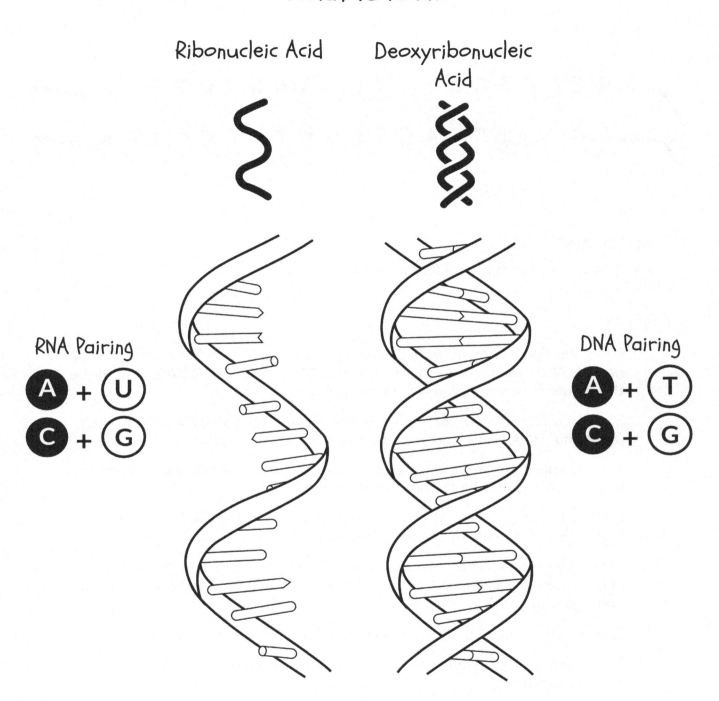

RNA Pairing

A + U

C + G

DNA Pairing

A + T

C + G

Remember the candy activity when DNA pairs A&T, C&G? RNA has uracil (instead of thymine) as one of the four nucleotides pairs.

ACTIVITY #7:
Make an RNA model with Pipe Cleaners

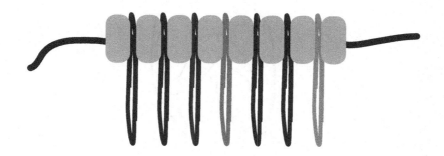

MATERIALS

— Pipe cleaners

— Paper clips with four different colors

— Plastic beads (large enough to go through the pipe cleaners)

STEPS

1. Sort out the paper clips by color. Then assign a color to each of the four RNA bases – adenine, uracil, cytosine, and guanine.

2. Hook a paper clip onto a plastic bead.

3. Thread the bead into the pipe cleaner to attach the RNA bases to the backbone.

4. Repeat as many times as you want.

Ebola is part of a handful of RNA viruses. Poly and Taq meet Laura Avena, a virologist. She is studying the Ebola virus so she can learn more about how it works. Now let's learn more about viruses.

SHAPES OF VIRUSES

Viruses come in an amazing variety of shapes and sizes.

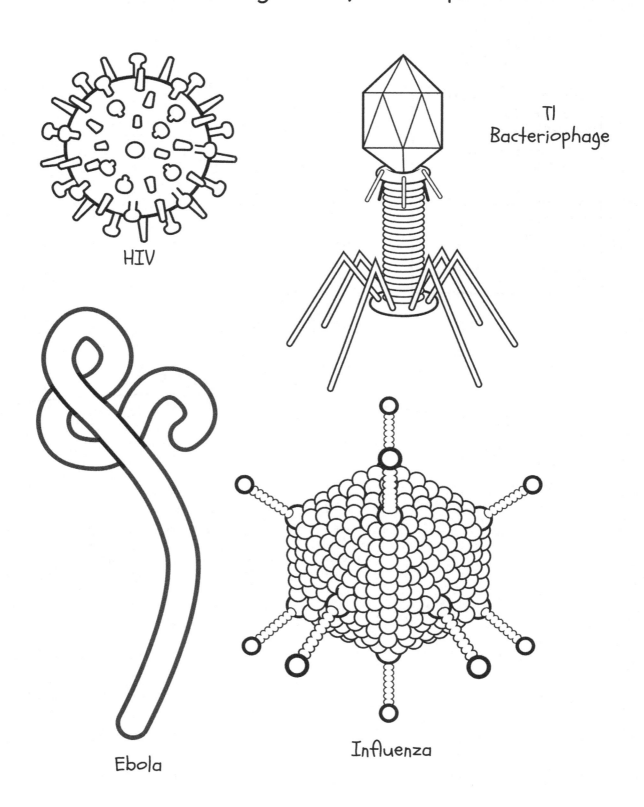

HIV

T1
Bacteriophage

Ebola

Influenza

CONNECT THE DOT:
Bacteriophage

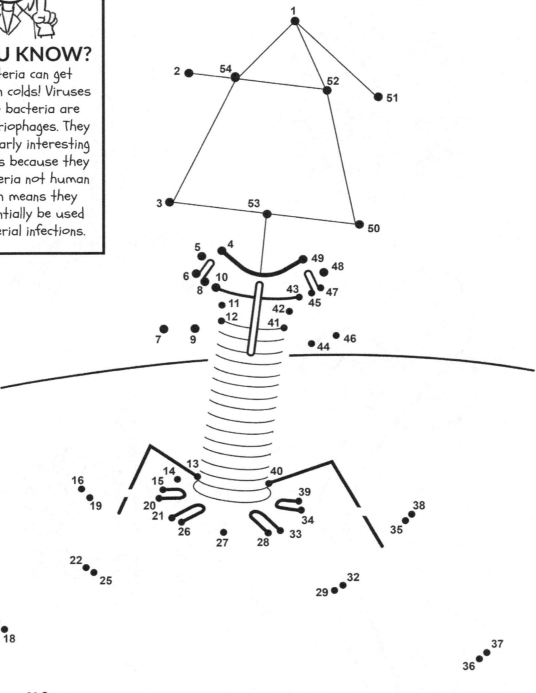

DID YOU KNOW?
Even bacteria can get infected with colds! Viruses that infect bacteria are called bacteriophages. They are particularly interesting to scientists because they attack bacteria not human cells which means they could potentially be used treat bacterial infections.

ANSWER KEY IS ON PAGE 120

ACTIVITY #8:
Make Your Own Bacteriophage Model at Home

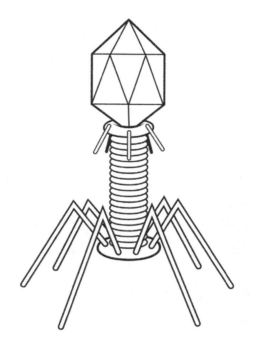

MATERIALS

- Bacteriophage Cut Out (See Page 131)
- Scissors
- Glue or Glue Dots®

STEPS

1. With the help of an adult, cut out the body parts of the bacteriophage. Make sure you cut on the solid lines.

2. Fold the paper on the dotted lines.

3. Assemble the head part together adding glue or a Glue Dot® as needed.

4. Roll the neck portion and add a dab of glue or add a Glue Dot® to create a long tube.

5. Fold along fold line and attach to base place where the "X" is.

6. Add a dab of glue or add Glue Dot® to bottom of base plate and add the legs.

7. Join head section to neck with additional glue.

8. Be careful to add glue along the glue tab.

Taq loves skateboarding.

He decides to go skateboarding with his friends but he doesn't hear his friends say "watch out" and falls and injures his leg.

After his fall, Taq visits the doctor who tells him that he
has a fracture.

Can you spot where Taq broke his ankle? Color the bone he broke in red.

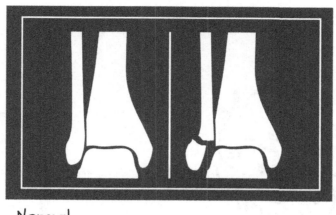

Normal Bone   Transverse Fracture   Oblique Fracture

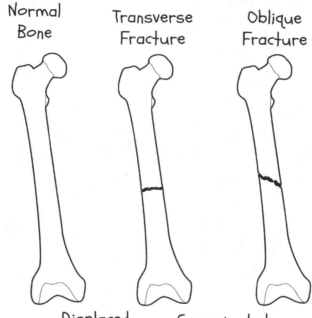

Displaced Fracture   Comminuted Fracture

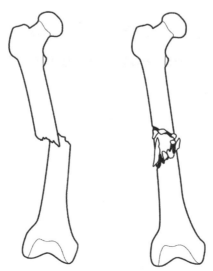

Taq asks Ms. Salinas if it's common for people to break their bones. Ms. Salinas says she knows the perfect person to ask.

Dr. Rama Sharma researches osteoporosis, or low bone mass. This affects almost 60 million people in the U.S. and leads to around nine million fractures annually—that's one fracture every three seconds!

Dr. Sharma explains that we attain our peak bone mass in our 20's that then drops continuously as we age. By age 50, one in three women and one in five men will experience a bone fracture.

FUN FACT

Astronauts lose on average 1% of bone mass per month spent in space.

Dr. Sharma recommends a few ways to keep your bones healthy. Can you circle the one that doesn't fit?

Swimming

Drinking milk
(especially fortified
with Vitamin D)

Running

Sitting for a
long time

Normal Skeleton

Skull

Clavicle (collar bone)

Rib

Humerus

Vertebra

Radius

Ulna

Pelvis

Femur

Patella

Tibia

Fibula

Which bones are missing?

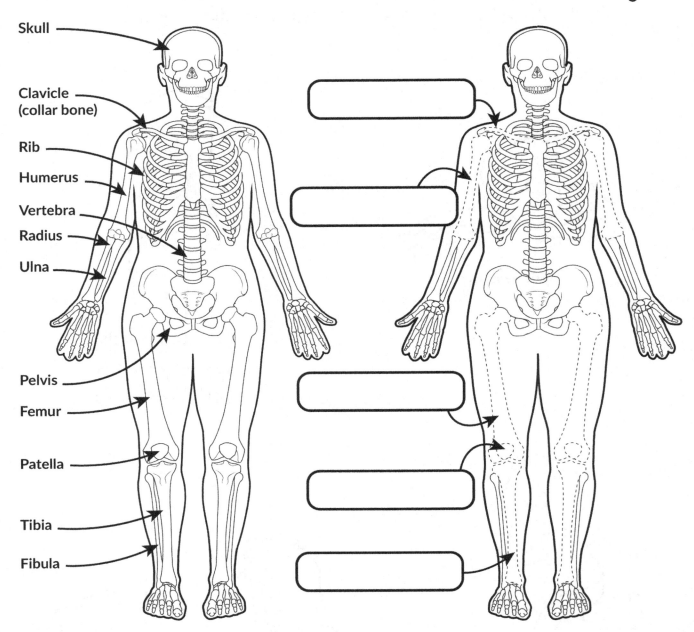

There are _____ bones in the human body.

Bones support your body and also protect your _____ .

Your _____ work with your bones to help you move around.

The longest bone in your body is the _____ .

A _____ is where two bones meet.

_____ helps keep your bones strong!

The smallest bone in your body is the _____ which is around 3 millimeters long!

WORD BANK

Joint
Organs
Muscles
206
Stapes
Femur
Calcium

ANSWER KEY IS ON PAGE 120

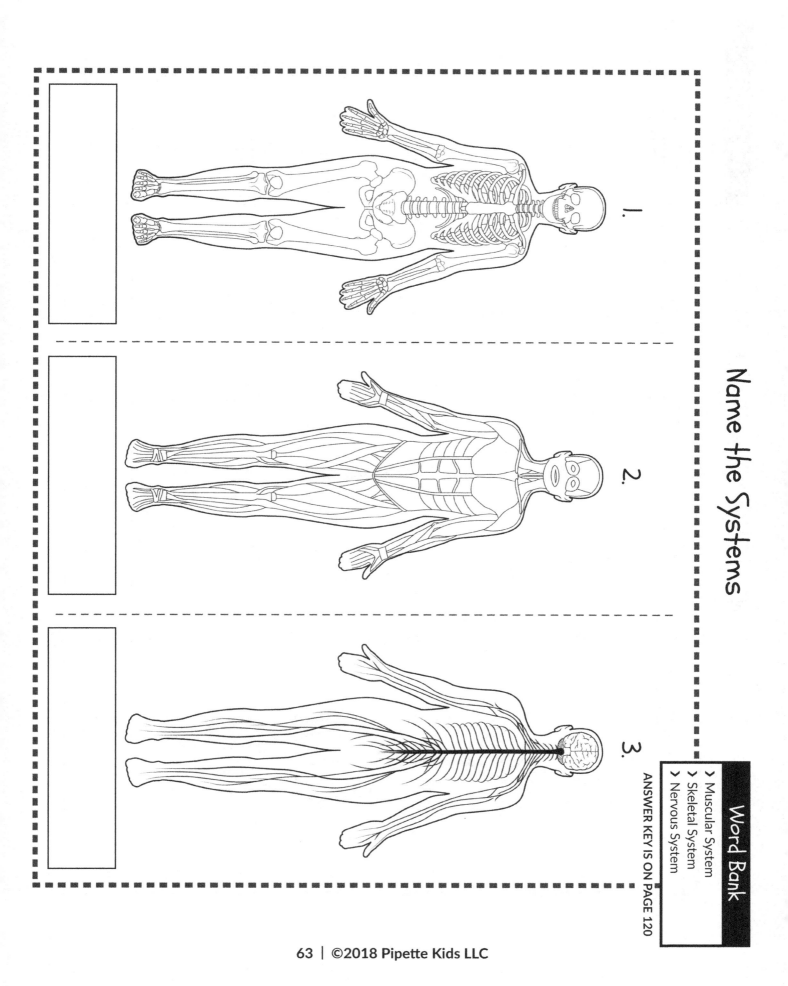

Name the Systems

1.

2.

3.

Name the Systems II

ANSWER KEY IS ON PAGE 121

Word Bank
> Respiratory System
> Cardiovascular System
> Digestive System

6.

5.

4.

CONNECT THE DOTS

What organs are inside your body? Connect the dots in order from 1 to 43.
Can you see the lungs, heart, liver, and stomach?

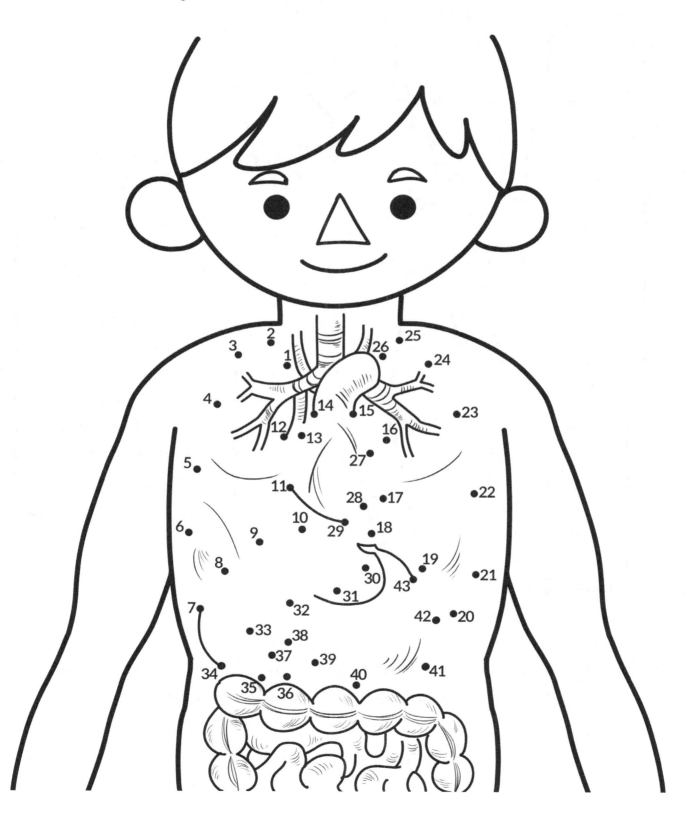

DID YOU KNOW?
Neuroscience is the study of the nervous system, including the brain, spinal cord, and networks of sensory nerve cells called neurons.

Dr. Santiago Ramon y Cajal is considered the father of neuroscience. He drew a lot of cells in the nervous system.

Nerve cells, called neurons, work together all over your body and send information to and from your brain. It's like a superhighway of messaging! These electrical signals are called impulses.

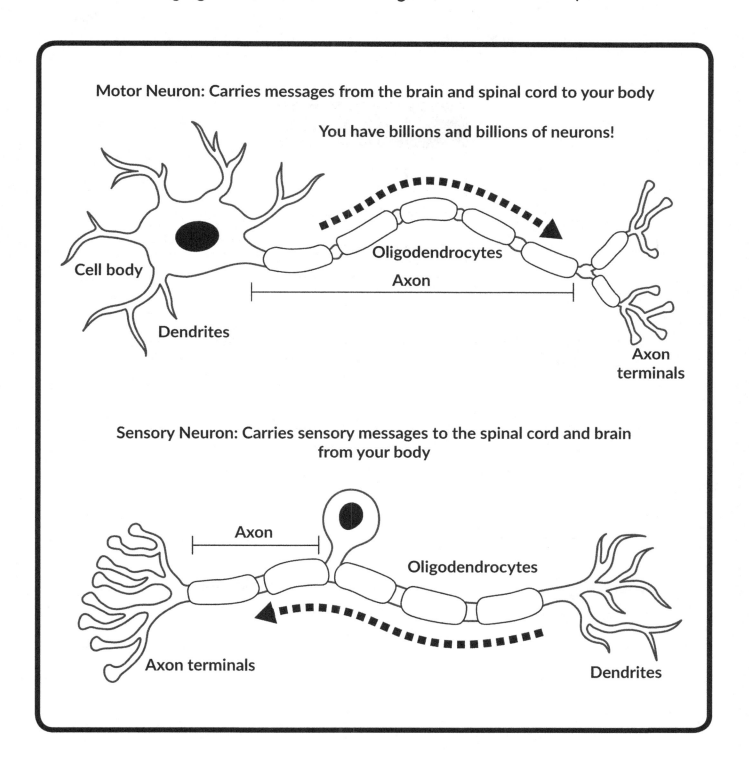

Motor Neuron: Carries messages from the brain and spinal cord to your body

You have billions and billions of neurons!

Cell body

Oligodendrocytes

Axon

Dendrites

Axon terminals

Sensory Neuron: Carries sensory messages to the spinal cord and brain from your body

Axon

Oligodendrocytes

Axon terminals

Dendrites

WORD BANK

Impulse

Motor neuron

Sensory neuron

When you want to move your arm, which type of nerve cell will help carry a message from your brain to your arm to make it move?

_____

Which type of nerve cell has a shorter axon?

_____

A nerve _____ is an electrical signal that travels along a nerve cell.

ANSWER KEY IS ON PAGE 121

Nerves send signals through the body. When you touch something
hot, your sensory neurons tell your brain that directs motor
neurons to pull your hand away.

|

ELECTRICAL SIGNALS

Your brain communicates with the rest of your body using tiny jolts of electricity. Can you trace the signal through the brain and down the brainstem? Don't run into the walls!

ANSWER KEY IS ON PAGE 121

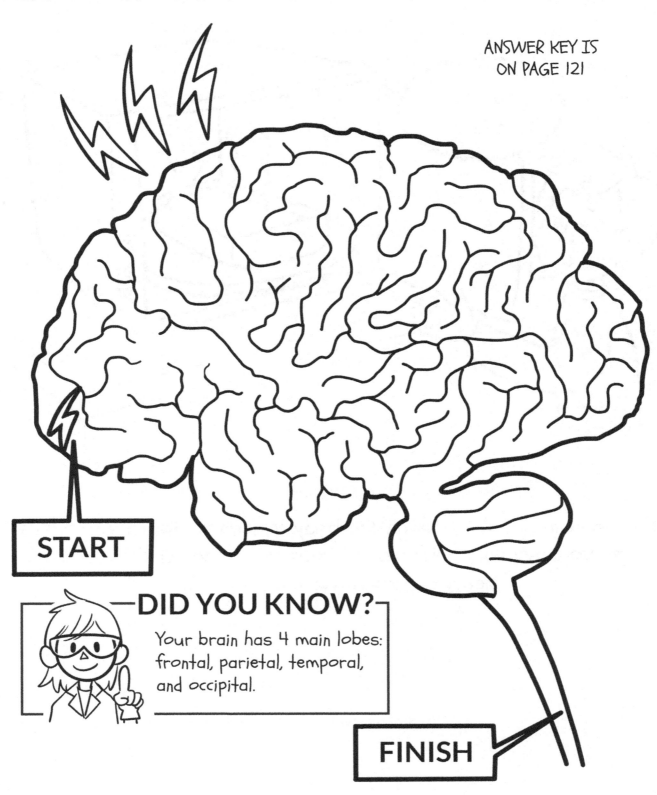

START

DID YOU KNOW?
Your brain has 4 main lobes: frontal, parietal, temporal, and occipital.

FINISH

This is Lily, a neuroscientist. Poly wants to know more about how her brain works and why we have memories. Lily tells them that our hippocampus helps turn short term memories (things we just learned) into long term memories (things you won't forget).

COLOR THE BRAIN

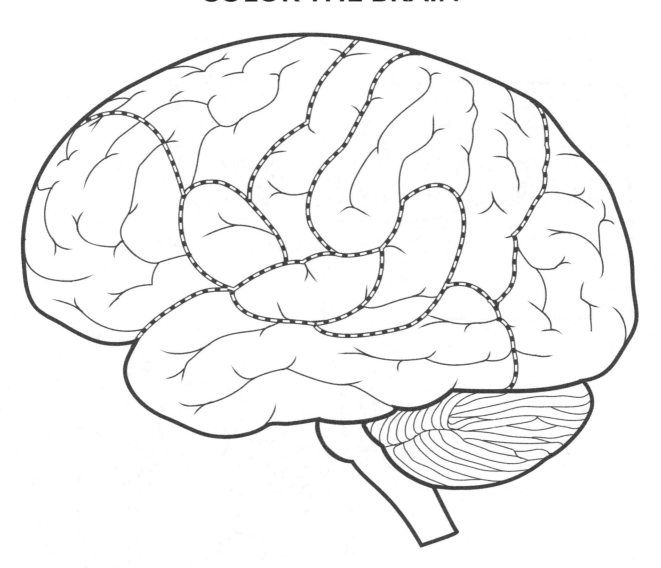

DID YOU KNOW?

Your brain is 73% water? It takes only 2% dehydration to affect your attention, memory, and other cognitive skills.

What happens when we eat too much junk food? Taq loves pizza and wants to know what happens to your body when you eat too many chips. He meets Allison Hester, a scientist that knows a lot about nutrition.

COLOR your PLATE

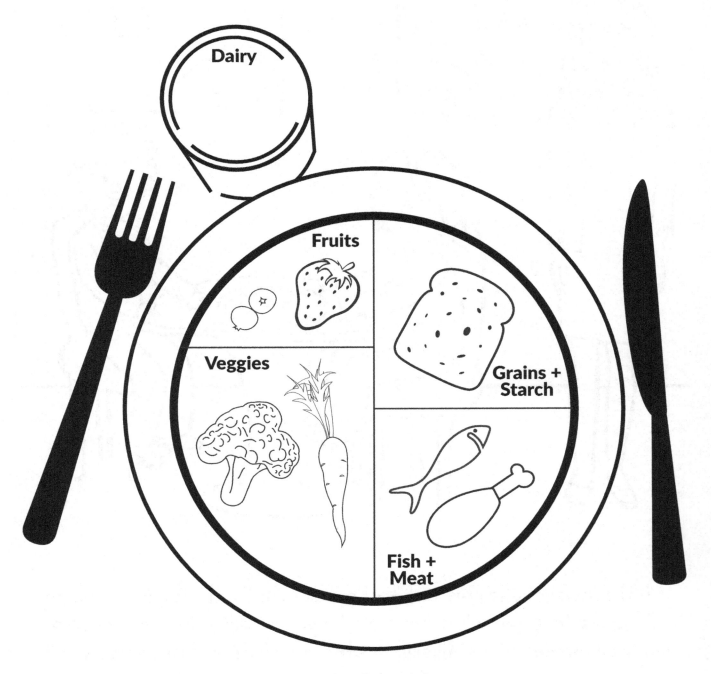

COLOR KEY

Fruits = Blue   **Grains + Starch = Yellow**

Fish + Meat = Red   **Veggies = Green**

|

THE DIGESTIVE SYSTEM

How it works:

Mouth: Teeth grind up food and salivary enzymes begin digestion to form a bolus

Esophagus: Has wave-like movements to help push bolus from the mouth to the stomach

Stomach: Bolus is further digested with new enzymes and gastic acid to form chyme

Small intestine: A long, winding tube where fluids from the pancreas and liver break down bolus from stomach even more, plus absorb nutrients from food

Large intestine: Absorbs water and minerals, and solid waste is created

Liver: Large organ that produces bile, and filters the blood

Gallbladder: Small, pear-shaped organ that stores bile until needed for digestion

Pancreas: Lies behind the stomach. Produces enzymes for digestion

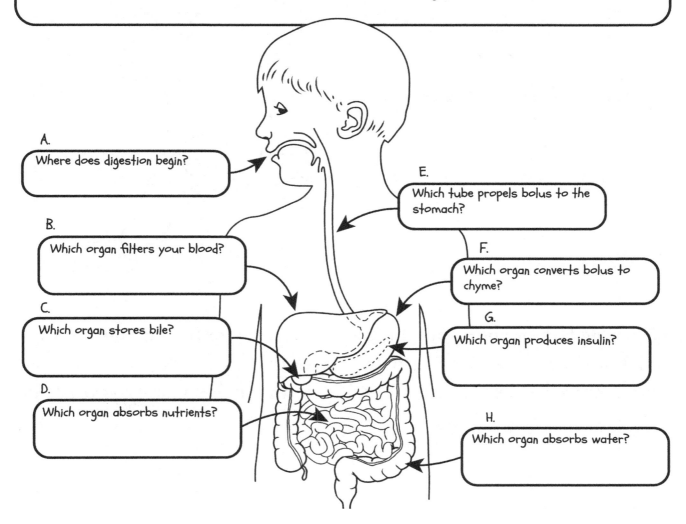

A. Where does digestion begin?

B. Which organ filters your blood?

C. Which organ stores bile?

D. Which organ absorbs nutrients?

E. Which tube propels bolus to the stomach?

F. Which organ converts bolus to chyme?

G. Which organ produces insulin?

H. Which organ absorbs water?

ANSWER KEY IS ON PAGE 121

Allison tells him that your body has to work extra hard when you eat food that has too much salt or sugar. Your body loves fruits and vegetables because it provides you with the energy you need.

But when you eat lots of candy and chips, your body gets a rush of energy and then crashes. Over time, if you eat too much salt and sugar your body won't be at its best because it is constantly working to keep up.

If you eat too much candy, you may get a cavity (a small hole in your teeth caused by bacteria). Your dentist will tell you to brush and floss regularly to avoid getting cavities.

MORE ABOUT YOUR TEETH

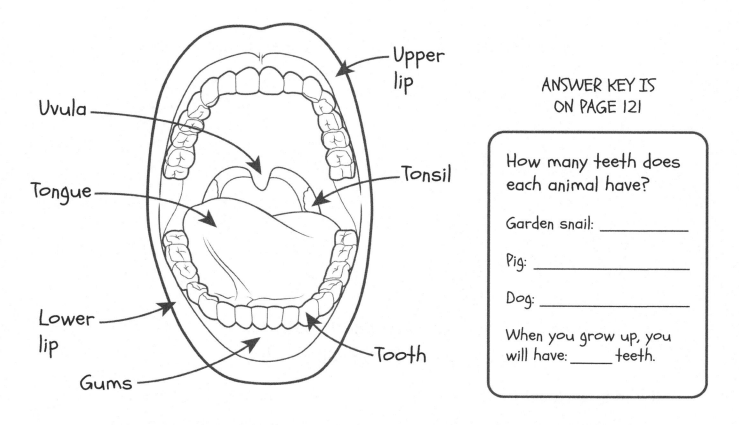

Upper lip

Uvula

Tonsil

Tongue

Lower lip

Tooth

Gums

ANSWER KEY IS ON PAGE 121

How many teeth does each animal have?

Garden snail: _____

Pig: _____

Dog: _____

When you grow up, you will have: _____ teeth.

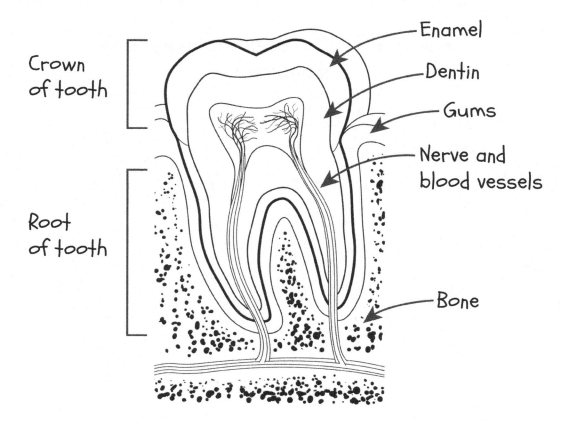

Crown of tooth

Root of tooth

Enamel

Dentin

Gums

Nerve and blood vessels

Bone

TEETH ACTIVITY

Find and circle the differences between these teeth!

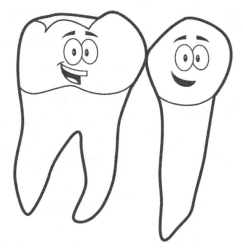 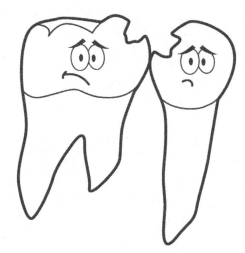

When you have all your adult teeth, how many of each type of tooth will you have?

I = Incisors (cuts food):   **C** = Canines (tears food):

P = Premolars (grinds food):   **M** = Molars (chews food):

Color each type of tooth a different color!

UPPER TEETH

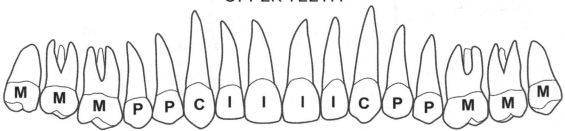

Back of mouth   Front of mouth   Back of mouth

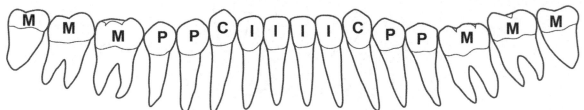

LOWER TEETH

In addition to eating well, keeping your heart healthy is important to maintaining a good body. In the 17th century, English physician William Harvey explained blood circulation for the first time showing there is a complete circuit that begins and ends in the heart.

Today, heart disease is one of the leading causes of death. In a matter of hours after a heart attack, the heart can lose ¼ of its function. Researchers like Ngoni are working to find out how to regrow damaged hearts. Unlike other organs, the adult heart has little to no capacity to heal itself once it has been damaged.

COLOR BY NUMBERS

1 Pink   Lungs take oxygen we breathe and give it to the blood, and release carbon dioxide when you exhale.

2 Blue   Blood that contains carbon dioxide, which is carried to lungs to pick up oxygen and drop off carbon dioxide.

3 Red   Blood that contains oxygen, travels all over the body to give oxygen to our cells.

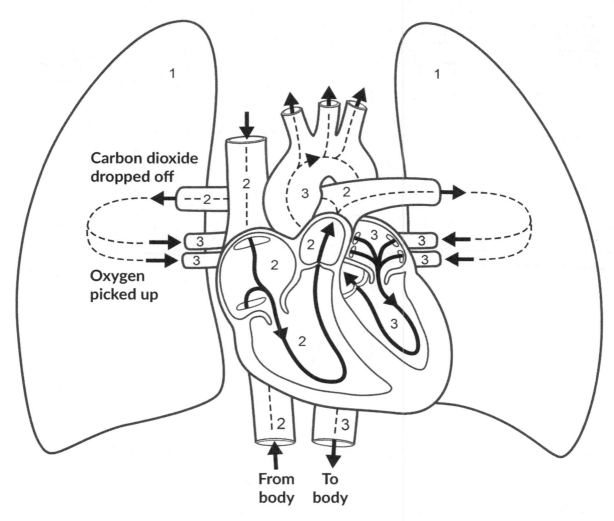

Follow the arrows to see how your blood travels through the chambers in your heart, to the lungs, back to the heart, and then to your body!

The heart transports blood to the different parts of the body. Blood has many things including white and red blood cells and other proteins. When you are hurt, your body can sense it and immediately sends red blood cells and proteins to plug the injury by clumping together. We call this a blood clot.

WHAT'S IN YOUR BLOOD?

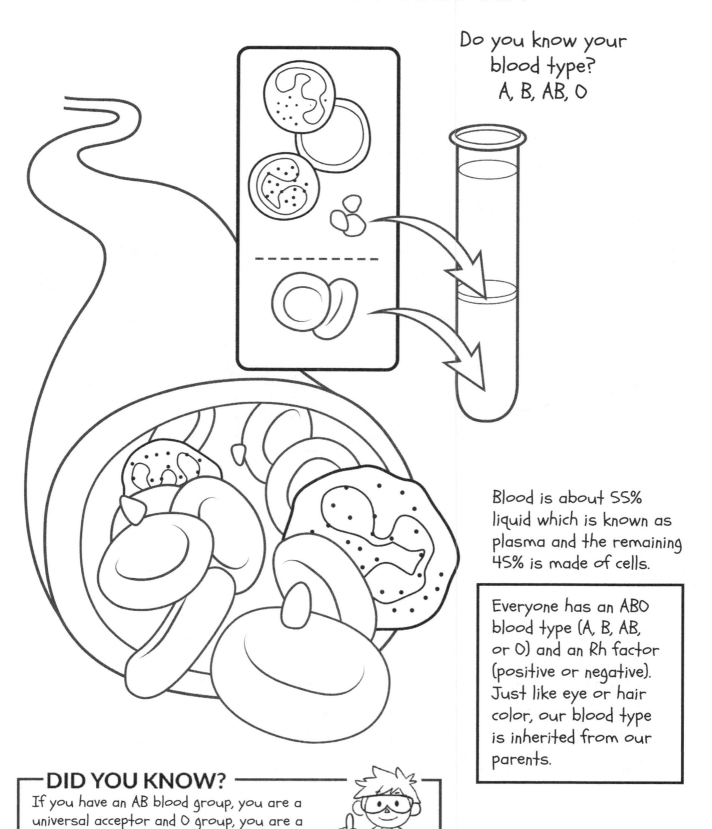

Do you know your blood type?
A, B, AB, O

Blood is about 55% liquid which is known as plasma and the remaining 45% is made of cells.

Everyone has an ABO blood type (A, B, AB, or O) and an Rh factor (positive or negative). Just like eye or hair color, our blood type is inherited from our parents.

DID YOU KNOW?

If you have an AB blood group, you are a universal acceptor and O group, you are a universal donor.

Taq wants to learn more about leukemia, the type of blood cancer that his grandma has. He meets Nema, a medical physicist who works at a cancer center. Nema shows Taq the linear accelerator, which is a machine that produces x-rays to zap cells in tumors.

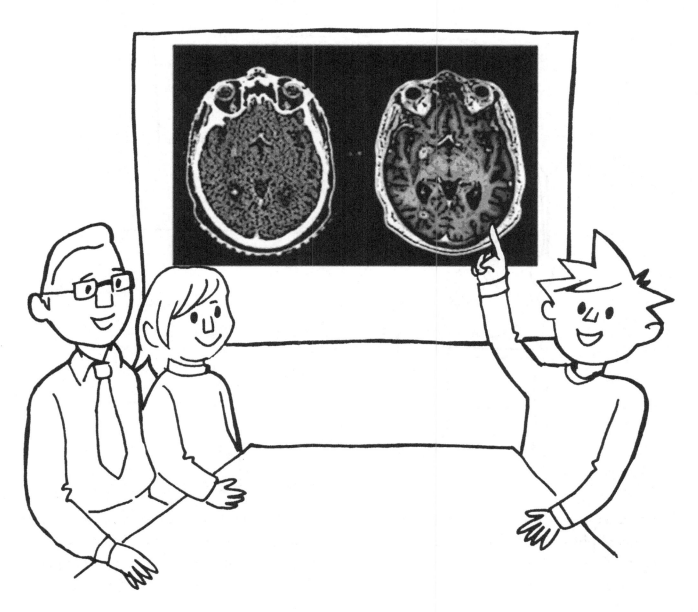

Nema explains that radiation is a form of energy that moves from one place to another. Radio waves, light, and x rays are examples of radiation.

Nema explains that the Periodic Table of Elements is very important in his work as a medical physicist. Russian chemist Dmitri Mendeleev organized the first draft of the Periodic Table of Elements and it is used by chemists to look at an element's properties at a glance. We know of 118 elements today and there may even be more elements we don't know about.

┌ DID YOU KNOW? ┐

Supposedly, Dmitri had a dream of a table where all the elements fell into place. He woke up immediately to write it down on a piece of paper.

A lot of the elements are already words you may know like Carbon, Oxygen, Gold, Copper or Nickel. One interesting element is Francium which was discovered by French physicist Marguerite Perey, one of the last chemical elements to be found naturally. Since then, a lot of elements have been produced artificially.

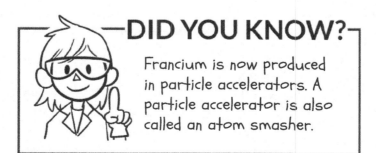

DID YOU KNOW?

Francium is now produced in particle accelerators. A particle accelerator is also called an atom smasher.

ELEMENTARY MYSTERY

Scientists use the periodic table to understand the characteristics of an element. Use the chart and the clues in each bubble to help you figure out each mystery element.

1	Atomic Number
H	Symbol
HYDROGEN	Name
1	Atomic Weight

8	13	20	11
O	**Al**	**Ca**	**Na**
OXYGEN	ALUMINUM	CALCIUM	SODIUM
1	27	40	23

15	29	16	1
P	**Cu**	**S**	**H**
PHOSPHORUS	COPPER	SULFUR	HYDROGEN
31	63.5	32	1

Forms table salt when combined with chloride

Element name:

The atomic number is 11

A shiny silver metal

Element name:

The atomic number is 13

You breathe with this gas

Element name:

Element symbol: O

Pennies are made from this metal

Element name:

Atomic weight: 63.5

ANSWER KEY IS ON PAGE 121

NAME THE ELEMENT

Here are some items you already know.

 Krypton

 Potassium

 Tin

 Chlorine

 Calcium

 Neon

 Sodium

 Strontium

 Carbon

 Lithium

 Helium

 Terbium

 Gold

 Lanthanum

 Beryllium

 Scandium

DID YOU KNOW?

The top 4 elements found in the human body are Oxygen, Carbon, Hydrogen, and Nitrogen.

Finish the Element Square

The Periodic Table contains the following information for each element.

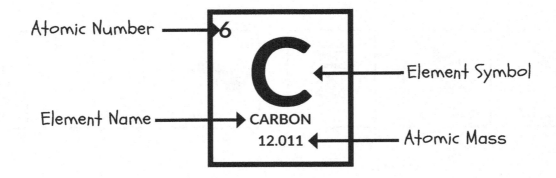

Pr	Gallum	16	Samarium	**Dy**
192.217	49	Radium	**Cf**	131.293
54	204.3833	**Cm**	74.9216	Polonium
Mercury	**Ni**	74	Rubidium	39.0983

FILL IN THE BLANK PERIODIC TABLE

EXAMPLE

6
C
CARBON
12.011

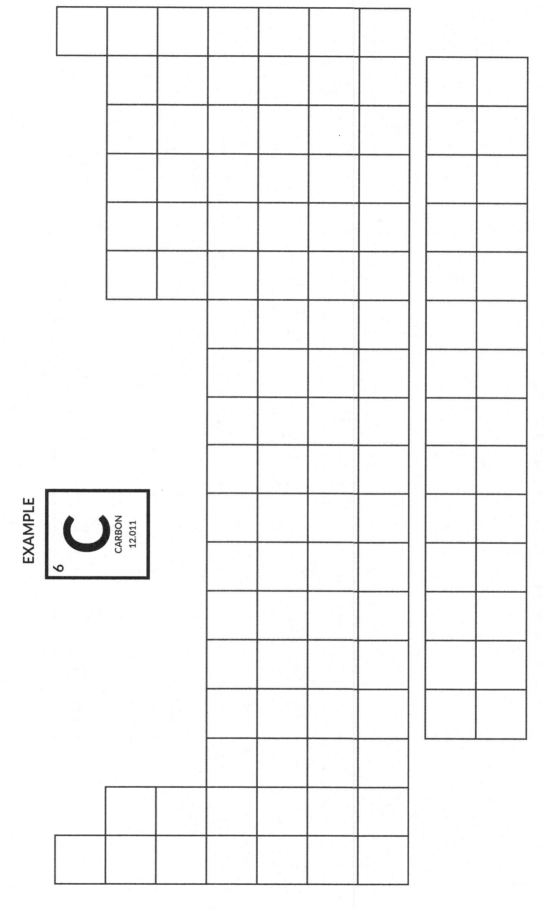

THE PERIODIC TABLE

1																	18
1 H Hydrogen																	2 He Helium
3 Li Lithium	4 Be Beryllium											5 B Boron	6 C Carbon	7 N Nitrogen	8 O Oxygen	9 F Fluorine	10 Ne Neon
11 Na Sodium	12 Mg Magnesium											13 Al Aluminium	14 Si Silicon	15 P Phosphorous	16 S Sulfur	17 Cl Chlorine	18 Ar Argon
19 K Potassium	20 Ca Calcium	21 Sc Scandium	22 Ti Titanium	23 V Vanadium	24 Cr Chromium	25 Mn Manganese	26 Fe Iron	27 Co Cobalt	28 Ni Nickel	29 Cu Copper	30 Zn Zinc	31 Ga Gallium	32 Ge Germanium	33 As Arsenic	34 Se Selenium	35 Br Bromine	36 Kr Krypton
37 Rb Rubidium	38 Sr Strontium	39 Y Yttrium	40 Zr Zirconium	41 Nb Niobium	42 Mo Molybdenum	43 Tc Technetium	44 Ru Ruthenium	45 Rh Rhodium	46 Pd Palladium	47 Ag Silver	48 Cd Cadmium	49 In Indium	50 Sn Tin	51 Sb Antimony	52 Te Tellurium	53 I Iodine	54 Xe Xenon
55 Cs Caesium	56 Ba Barium	57 La Lanthanum	72 Hf Hafnium	73 Ta Tantalum	74 W Tungsten	75 Re Rhenium	76 Os Osmium	77 Ir Iridium	78 Pt Platinum	79 Au Gold	80 Hg Mercury	81 Tl Thallium	82 Pb Lead	83 Bi Bismuth	84 Po Polonium	85 At Astatine	86 Rn Radon
87 Fr Francium	88 Ra Radium	89 Ac Actinium	104 Rf Rutherfordium	105 Db Dubnium	106 Sg Seaborgium	107 Bh Bohrium	108 Hs Hassium	109 Mt Meitnerium	110 Ds Darmstadtium	111 Rg Roentgenium	112 UUb Ununbium	113 UUt Ununtrium	114 UUq Ununquadium	115 UUp Ununpentium	116 UUh Ununhexium	117 UUs Ununseptium	118 UUo Ununoctium

58 Ce Cerium	59 Pr Praseodymium	60 Nd Neodymium	61 Pm Promethium	62 Sm Samarium	63 Eu Europium	64 Gd Gadolinium	65 Tb Terbium	66 Dy Dysprosium	67 Ho Holmium	68 Er Erbium	69 Tm Thulium	70 Yb Ytterbium	71 Lu Lutetium
90 Th Thorium	91 Pa Protactinium	92 U Uranium	93 Np Neptunium	94 Pu Plutonium	95 Am Americium	96 Cm Curium	97 Bk Berkelium	98 Cf Californium	99 Es Einsteinium	100 Fm Fermium	101 Md Mendelevium	102 No Nobelium	103 Lr Lawrencium

Poly has always wanted to meet a biomedical engineer. Ms. Salinas introduces Poly and Taq to Sharon, a concert violinist and a biomedical engineer. Sharon tells Poly about an invention she is working on. Sharon explains that one day when she was playing her violin outside, she realized that it didn't sound very good. She tried her rosin, which is a block that helps her strings sound better. She then realized that her rosin did not do well in humid areas like San Antonio.

Putting on her science goggles, Sharon went to the lab and started to research other ingredients which would work better. She is currently tinkering with different recipes and hopes to find the "perfect" rosin.

WHO'S WHO

One of the main jobs of a scientist is to discover new things and tell the world about their findings. Look at the pictures of each scientist. Using clues from the picture, match the scientist to the description about their discoveries. The answer key is on page 115.

RACHEL CARSON

An American marine biologist, conservationist, and writer. She famously wrote a book called Silent Spring, about the dangers of using pesticides.

MARIA S. MERIAN

A German entomologist, naturalist, explorer, and artist. She was the first person to document the life cycle of insects with her meticulous illustrations.

CAROLINE HERSCHEL

A German astronomer. She discovered many comets and was the first woman to be an honorary member of the Royal Astronomical Society.

ROSALIND FRANKLIN

An English physical chemist and X-ray crystallographer. She was the first to discover the image of DNA using X-ray crystallography. She laid the groundwork for many Nobel laureates.

INVENTORS

Life Raft
Maria Beasely
1880

Dishwasher
Josephine Cochrane
1886

Square Bottomed Paper Bags
Margaret Knight
1870

Refrigerator
Florence Parpart
1914

Windshield Wipers
Mary Anderson
1903

Fire Escape Bridge
Anna Connelly
1887

A PATH TO BECOME A SCIENTIST (OR MAKE YOUR OWN!)

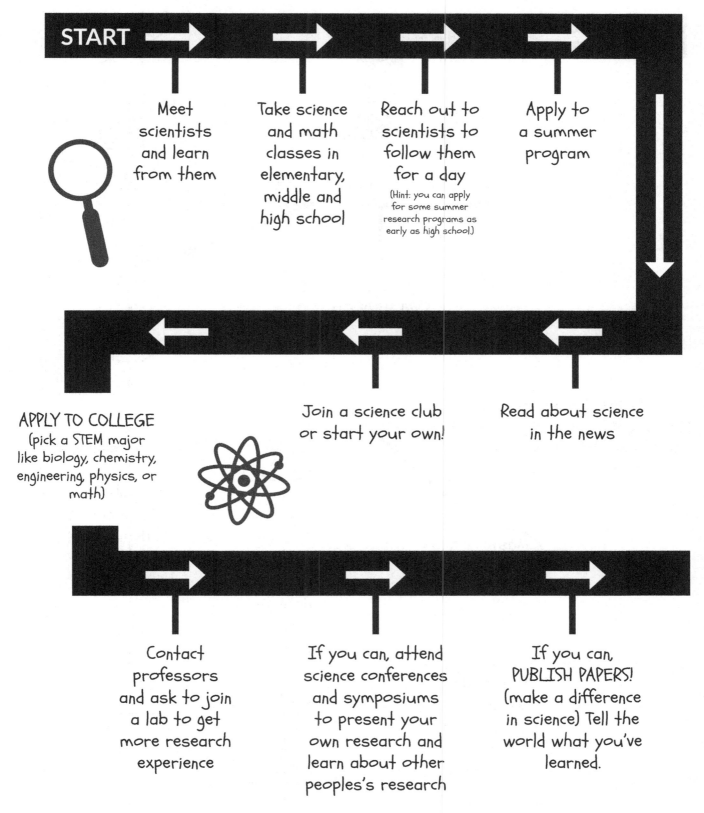

START →

Meet scientists and learn from them

Take science and math classes in elementary, middle and high school

Reach out to scientists to follow them for a day
(Hint: you can apply for some summer research programs as early as high school.)

Apply to a summer program

Read about science in the news

Join a science club or start your own!

APPLY TO COLLEGE (pick a STEM major like biology, chemistry, engineering, physics, or math)

Contact professors and ask to join a lab to get more research experience

If you can, attend science conferences and symposiums to present your own research and learn about other peoples's research

If you can, PUBLISH PAPERS! (make a difference in science) Tell the world what you've learned.

PATH TO BECOME A SCIENTIST

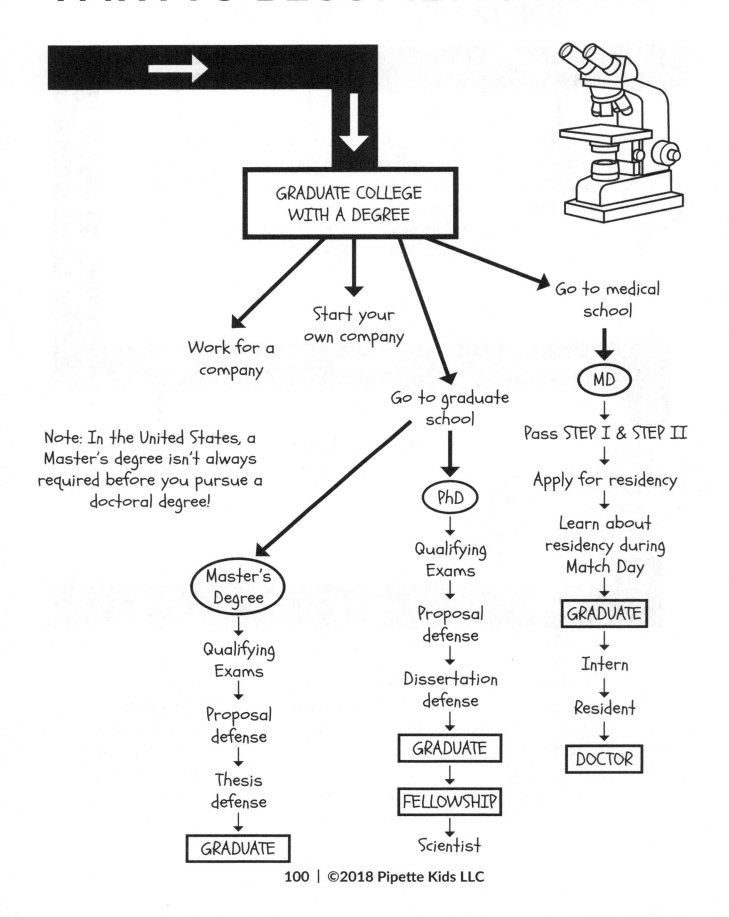

GRADUATE COLLEGE WITH A DEGREE

Work for a company

Start your own company

Go to graduate school

Go to medical school

Note: In the United States, a Master's degree isn't always required before you pursue a doctoral degree!

Master's Degree
↓
Qualifying Exams
↓
Proposal defense
↓
Thesis defense
↓
GRADUATE

PhD
↓
Qualifying Exams
↓
Proposal defense
↓
Dissertation defense
↓
GRADUATE
↓
FELLOWSHIP
↓
Scientist

MD
↓
Pass STEP I & STEP II
↓
Apply for residency
↓
Learn about residency during Match Day
↓
GRADUATE
↓
Intern
↓
Resident
↓
DOCTOR

CAREERS IN SCIENCE
Complete the crossword

ANSWER KEY IS
ON PAGE 121

WORD BANK
> PHYSIOLOGIST
> CYTOLOGIST
> RADIOLOGIST
> DENTAL SCIENTIST
> TOXICOLOGIST
> MOLECULAR BIOLOGIST
> PHARMACOLOGIST
> BIOMEDICAL ENGINEER
> CANCER BIOLOGIST
> BIOLOGIST
> PHYSICIST
> MICROBIOLOGIST
> GERONTOLOGIST
> TRANSLATIONAL SCIENTIST
> BIOCHEMIST
> VIROLOGIST
> IMMUNOLOGIST
> NURSING SCIENTIST
> GENETICIST
> NEUROSCIENTIST

ACROSS
1. Studies motion, forces, and energy
8. Studies microorganisms such as bacteria, viruses, algae, fungi, and parasites
13. Studies how research in the lab can be used to treat patients
17. Studies the chemical properties and interactions of biological molecules
18. Studies viruses
19. Studies how traits are inherited
20. Studies the nervous system

DOWN
2. Studies the immune system
3. Studies how to improve the field of nursing
4. Studies the human body
5. Studies cells
6. Studies how to diagnose and treat diseases based on x-rays or nuclear radiation
7. Studies diseases related to the teeth, gums, and mouth
9. Studies the impact of toxic materials
10. Studies the molecular basis in the various systems of a cell
11. Studies how drugs work
12. Studies how to design and develop medical products
14. Studies cancers and tumors
15. Studies life forms and the environment
16. Studies the aging process

MEET THE RESEARCHERS

"I always wanted to know why things were happening. Research gives you the opportunity to do that and allows you to learn what's outside of the textbook. It's exciting to me."

Sabrina Martinez

"If I love doing medicine and I love science and discovery, then it hit me, why don't I just do both. As a clinician-scientist, I would have the ability to move in between both worlds. I could take care of patients but also use science to improve care for my patients."

Vinh Dao

"Growing up in Chicago, no one talked about pursuing a career in research. It's important for students to know at an early age that research is a rewarding and challenging possibility. It's also important for women of color to know that they can do great things in STEM fields."

Calais Prince

MEET THE RESEARCHERS

"I love coming in every day and working toward solving problems that will eventually help us understand nature and life a little better. The feeling of accomplishment of getting a result you didn't expect is like nothing I've felt in any other part of my life. I feel like Sherlock Holmes, but a lot cooler."

Phillip Webster

"I was enrolled in AP Biology, where we had to learn about cellular processes in great detail...I remember being so enthralled by what a cell is capable of doing, and the intricacies of the mechanisms behind its actions. Since then, I just wanted to learn more."

Alison Clark

"Curiosity and necessity are the founding stones for great discoveries. I had the curiosity, I found the necessity in Neuroscience. The human brain is not just an organ but a piece of art! This enigma is what I am passionate about and draws me to the laboratory each day. My mentor once told me, it is not so much the intelligence that helps make people discoveries but it's their motivation, character, and flexibility that allows them to make those discoveries!"

Mustafa Mithaiwala

MEET THE RESEARCHERS

"As a scientist, not only have I learned about science, but I have also learned from science. I have learned to never give up and to not fear failure. Failure can teach us much more about ourselves and about the problem than success can. And usually the harder the problem, the more important its solution is to the world."

Ahsan Choudary

"People told me that I couldn't be a dancer, ice skater, and a scientist at the same time. I wanted to prove them wrong, so I tried out for the Dallas Stars Ice Girls and made the team! Being a Dallas Stars Ice Girl while doing research was one of the best and most fun years I have ever experienced. Having interests and passions outside of the lab keeps me happy and healthy. When I am not in lab, I like to volunteer, work out, and go hiking. Exercise helps me think and feel better! Remember, you can be anything and anyone you want to be. There are no limits and rules."

Samantha Yee

"Science is constantly moving forward, and sometimes the complex questions get simpler with improvements in the field. So don't be afraid to ask 'silly' questions. The answers just might become the key to discovering your passion."

Allie Sharp

MEET THE RESEARCHERS

"Always keep learning because you should be aware of what is known so you can understand what is not known. Never stop learning."

Brian Iskra

"Don't be afraid to ask questions. Your curiosity could lead to the latest cutting-edge technology or even groundbreaking discoveries."

Cassidy Daw

"Don't let anyone tell you that you can't be a scientist. I'll let you in on a secret – throughout elementary, middle and high school, I never got top scores on math or science tests. My academic performance was just above average. But what I did do, was cultivate my creative abilities. I was passionate about dancing, painting, and acting. It's the balance of creativity, curiosity and perseverance that's put me on the path to becoming a scientist. Always remember, you can do what you set your mind to."

Roma Kaul

MEET THE RESEARCHERS

"Everyday, I am asked to be the very best version of myself as a biomedical scientist. Biomedical research is complicated and hard, but rewarding. Science moves medicine forward and the integration of the two makes lives better. Science is progress and its impact is global. And I am here for all of it."

Sadiya Ahmad

"Always be willing to challenge yourself. It's exciting to learn new things, even if they are hard sometimes. Like when you were a baby, you could not do a lot of things, but you learned how to because you tried hard and didn't give up. As you grow up, don't be scared of a challenge. Study science if you like it, even if other people don't like it or they say you can't. And whenever you mess up, don't worry, everybody does, even all the scientists in this book! Just be patient, be strong and keep trying."

Alex Kirkpatrick

Learn more at
www.PipetteKids.com

MEET THE ARTISTS

Sue Simon is the medical illustrator who drew all the bacteria, viruses, hearts, teeth, brains, and all the other body parts in this book. She has a master's degree and loves to learn new things! Sue likes to go camping, paint in watercolors, and volunteer.

Rebecca Osborne is a British illustrator with a not-so-secret passion for science. She loves to use her work to encourage and inspire the next generation of scientists (that's you!) Rebecca lives in an English seaside town with her family and two dogs.

Angela Gao is a scientist who draws things, too. Her favorite part about science is biology, especially the small things like bacteria and viruses. Her favorite part about art is cartoons and comic books.

THE 5 STEPS OF THE SCIENTIFIC METHOD

1. Make Observations!
Use your 5 senses: sight, hearing, taste, touch and smell, to learn about the world around you.

2. Ask Questions!
Scientific investigations begin with ideas that you're not sure about.

3. Hypothesize!
Make an educated guess, or hypothesis, that you can test.

4. Do an Experiment
Plan how to test your hypothesis- design an experiment.

5. Draw a Conclusion!
A conclusion is a statement that sums up what you learned.

THE SCIENTIFIC METHOD

1. List your observations:

_____

2. Question:

_____

3. Make a hypothesis:

_____

4. Design an experiment to test your hypothesis:

_____

5. What is your conclusion?

_____

Now it's time to do your own activities!

ACTIVITY #9:
Make Your Own Lava Bottle

MATERIALS

— A clear 1 liter soda bottle

— Measuring cup

— Funnel

— Water

— Vegetable oil

— Food coloring

— Seltzer tablets

STEPS

1. Pour the ¾ cup of water into the bottle.

2. Using the funnel, pour oil into the bottle until it's almost full.

3. Add a few drops of food coloring into the bottle.

4. Allow the water and the oil to separate?

5. To watch the lava bottle in action, add half of the seltzer tablet into the bottle.

ACTIVITY #10:
Grow Rock Candy

MATERIALS

- 3 cups of sugar
- 1 cup of water
- String
- Popsicle sticks
- 1 glass jar
- Food coloring (optional)
- Flavoring (optional)
- Saucepan

STEPS

1. Take some string and wrap it around the popsicle stick. Leave some string to hang (this is where your rock candy will grow).

2. Add the water and sugar into the saucepan. Stir and bring it to a boil with the help from an adult. Make sure that all the sugar crystals are dissolved.

3. Remove the sugar solution from the heat. At this point, you can add food coloring or flavoring to the solution.

4. Pour sugar solution into the glass jar.

5. Place the popsicle stick with the string on the mouth of the jar. Make sure that the string does not touch the bottom or the sides of the jar.

6. Cover the jar with a paper towel, and place the jar in an area where it will not be bumped or spilled.

7. Check your rock candy daily. It may take a day or two to begin seeing your rock candy grow. Remove crystals that are growing on top of the jar, since they will compete with the growth of your rock candy. Likewise, you can always transfer the rock candy and the sugar solution into a clean glass jar.

8. When you are happy with the size of your rock candy, remove it from the solution and allow to dry.

ACTIVITY #11:
Make Your Own Ice Cream

MATERIALS

- 2–3 cups ice, crushed or cubed
- 1/3 cup kosher or coarse salt
- 1 quart-sized plastic ziplock bag
- 1 gallon-sized plastic ziplock freezer bag
- 1 cup milk (can also be substituted with almond milk or rice milk)
- 1.5 tablespoons sugar
- 1/2 teaspoon vanilla

STEPS

1. Combine milk, sugar, and vanilla inside the quart ziplock bag to make the ice cream mixture. Seal the bag tightly, getting rid of as much air as you can.

2. Mix ice and salt in the gallon ziplock bag.

3. Place the bag with the ice cream mixture inside the bag of ice. Seal the bag tightly.

4. Shake the bag vigorously for about 5 minutes. You should see the ice cream mixture harden over time.

5. Allow the mixture to set on a plate or towel for an additional 5 minutes. Without opening the bag, arrange the ice so that the bag with the ice cream is surrounded by ice.

6. Take out the bag and scoop the ice cream. Serve immediately.

|

ACTIVITY #12:
Make Invisible Ink

MATERIALS

- 1 lemon, cut in half
- Water
- Spoon
- 1 small bowl
- Computer paper
- Cotton swabs
- Lamp

STEPS

1. Squeeze both halves of the lemon over the bowl.

2. Add 2-3 drops of water to dilute lemon juice and mix with your spoon.

3. Use a cotton swab as your writing utensil. Dip it in the juice and write your message on the computer paper.

4. To see your message, hold your paper close to the lamp.

ACTIVITY #13:
Tornado in a bottle

MATERIALS

- Water
- Clear plastic bottle with a cap
- Glitter
- Dishwashing soap

STEPS:

1. Fill the bottle with water until ¾ full.

2. Add a few drops of dishwashing liquid.

3. Take a couple pinches of glitter and add to the water.

4. Put the cap on the bottle tightly.

5. Turn the bottle upside down, and hold it by the neck. Spin the bottle in a circular motion for a few seconds.

6. Stop and see if you can see a mini tornado forming inside the bottle!

ACTIVITY #14:
Rainbow in a Glass

MATERIALS

— 6 glasses or cups

— 1 measuring cup

— 1 measuring spoon

— Food coloring

— Hot tap water

— Sugar

— A 16 oz glass mason jar

— Turkey baster

STEPS

1. Measure 1 cup of water and pour into each glass. Make sure each glass has the same amount of water.

2. Make your rainbow water using the food coloring! Add color to your water so that you have red, orange, yellow, green, blue, and purple water.

3. To make water with different densities, you will need to add different amounts to your rainbow water. Leave the red water as is (no sugar). Start by adding 2 teaspoons of sugar to your orange water. Double the amount of teaspoons as you work your way through the rainbow water (for example, add 4 tsp in yellow water, 6 tsp. in green, 8 tsp. in blue, 10 tsp. in purple). Stir until all the sugar is dissolved.

4. With your turkey baster, suck up some of the purple sugar water and squeeze into the mason jar. Make sure you keep the side of the glass dry by squeezing on to the middle.

5. Next, add the blue water. Very carefully add the blue water on the surface of the purple water, towards the middle.

6. Repeat the same steps until all the colors have been added. Do you have a rainbow in a jar?

ACTIVITY #15:
The Egg and the Bottle

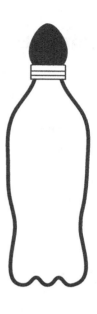

MATERIALS

- 1 peeled, boiled egg
- 1 glass bottle
- Paper strips
- Lighter or matches

STEPS

1. Do this experiment with adult supervision since there will be fire involved.

2. Set the egg over the mouth of the glass bottle. The egg should be slightly larger than the mouth of the bottle.

3. Set a strip of paper on fire using a lighter or a match. Lift the egg off the bottle, drop the paper inside the bottle, and very quickly replace the egg on the mouth of the bottle.

4. Observe as the egg gets sucked inside the bottle.

ACTIVITY #16:
Milk of Magnesia Experiment

MATERIALS

— Beet root juice

— Milk of Magnesia

— Water

— Vinegar

— 16 oz clear cup (can be plastic or glass)

— Popsicle sticks

— Measuring spoon

STEPS

1. Pour milk of magnesia in the cup until the cup is 1/4 of the way full. Add an equal amount of water to the cup.

2. Add about two tablespoons of beet root juice to the cup. Stir using a popsicle stick. What did you see happen?

3. While stirring the solution, add the vinegar one tablespoon at a time. Observe the color change.

FUN FACT

The color change reflects the change in acidity which is a measure of the number of hydrogen ions in the solution.

ACTIVITY #17:
Non-Newtonian Fluid

MATERIALS

- 1 box (1 lb) of cornstarch
- 1-2 cups of water
- 1 large mixing bowl or container
- 1 baking sheet (a baking tray with sides will work as well)
- 1 gallon ziplock bag
- Paper towels for clean up
- Food coloring if desired

STEPS

1. Put cornstarch in the mixing bowl.

2. Add 1 cup of water to the cornstarch and mix by hand. Make sure that your hands are clean!

3. Continue adding water until the consistency of the mixture is like honey.

4. Add food coloring if desired.

5. Transfer the mixture onto the baking sheet or tray. What do you notice about its consistency?

6. Stir the mixture with your fingers, first very slowly, and then as fast as you can. What do you observe?

7. Grab a handful of the mixture and make a ball using your hands. Hand over the ball to someone. What do you see happening?

8. When you are finished, pour the mixture inside the ziplock bag to store it.

ANSWER KEY

PAGE 11: CONNECT THE DOTS: MICROSCOPE

PAGE 12:
1. Butterfly Wing 2. Sand
3. Velcro 4. Strawberry

PAGE 23: TOOLS OF A SCIENTIST

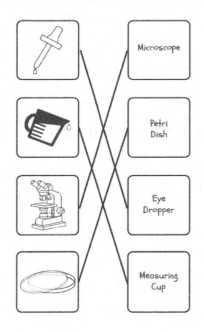

PAGE 25: SO MANY TOOLS
1. Bunsen Burner 2. Pipette 3. Microscope Slides
4. Petri Dish 5. Erlenmeyer Flask 6. Test Tube Rack
7. Funnel 8. Goggles 9. Incubator 10. Gel Boxes
11. Gloves 12. Weighing Balance 13. Beaker
14. Centrifuge

PAGE 26: 9

PAGE 27:

You did a
GREAT JOB!

PAGE 28:
1. Centrifuge 2. Pipette 3. Incubator

PAGE 29:
A. 4 B. 0 C. 13 D. 2

PAGE 35: CELL CITY
5: Mitochondria 1: Vacuole
2: Centrioles 1: Nucleus 2: Golgi
complexes 1: Endoplasmic reticulum
6: Ribosomes 8: Lysosomes

PAGE 52:

PAGE 62:
1. 206 2. Organs 3. Muscles 4. Femur
5. Joint 6. Calcium 7. Stapes

PAGE 63:
1. Skeletal System 2. Muscular System
3. Nervous System

ANSWER KEY

PAGE 64:
1. Cardiovascular System
2. Respiratory System
3. Digestive System

PAGE 68:
1. Motor neuron 2. Sensory neuron 3. Impulse

PAGE 70: Electrical Signals

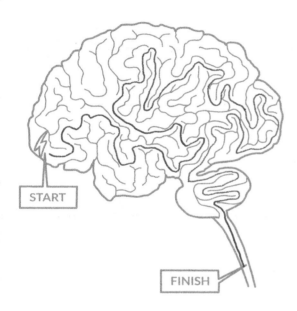

PAGE 75:
A. Mouth B. Liver C. Gall Bladder D. Small Intestine
E. Esophagus F. Stomach G. Pancreas H. Large
Intestine

PAGE 79:
1. 14, 000 2. 28 3. 42 4. 32

PAGE 90: Elementary Mystery
SODIUM forms table salt when combined with chloride.
ALUMINUM is a shiny silver metal.
OXYGEN is a gas in the air that you breathe.
COPPER is the metal the pennies are made from.

PAGE 97: Who's Who?

Rosalind Franklin   Maria S. Merian

Caroline Herschel   Rachel Carson

PAGE 101:
1. Physicist 2. Immunologist 3. Nursing Scientist 4. Physiologist 5. Cytologist 6. Radiologist 7. Dental Scientist 8. Microbiologist 9. Toxicologist 10. Molecular Biologist 11. Pharmacologist 12. Biomedical Engineer 13. Translational Scientist 14. Cancer Biologist 15. Biologist 16. Gerontologist 17. Biochemist 18. Virologist 19. Geneticist 20. Neuroscientist

Now help Poly and Taq write their report.
Which scientist or inventor inspires you?

_____

_____

_____

_____

_____

_____

_____

_____

_____

_____

_____

_____

Now help Poly and Taq write their report.
Which scientist or inventor inspires you?

_____

_____

_____

_____

_____

_____

_____

_____

_____

_____

|

BRAIN HAT INSTRUCTIONS

Cut out both hemispheres

Tuck in flaps and tape

Wear your new brain hat!

|

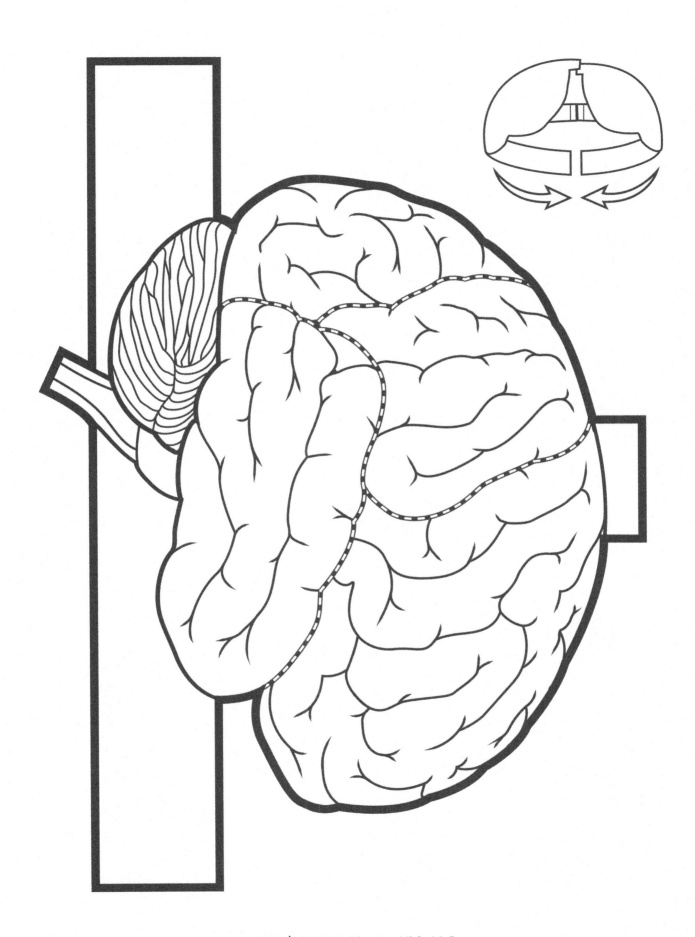

|

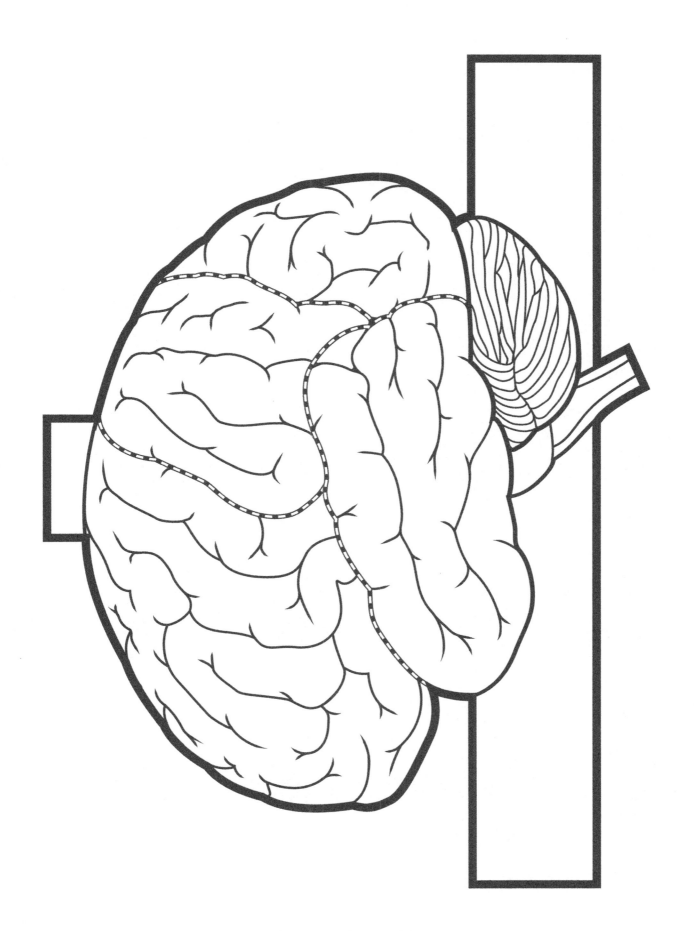

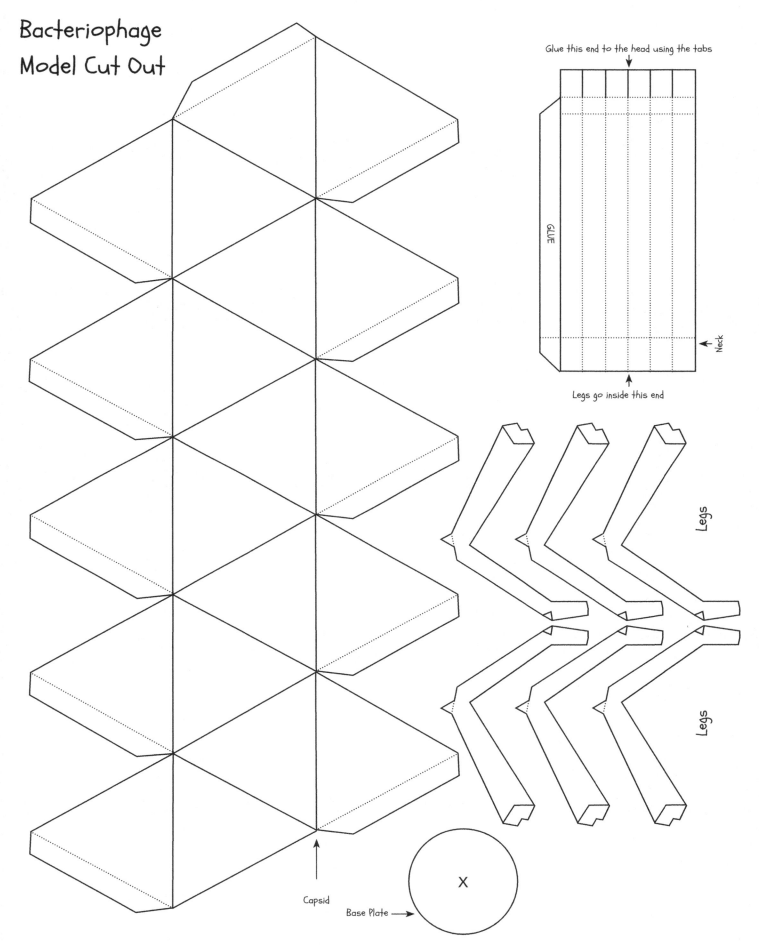

Bacteriophage Model Cut Out

Glue this end to the head using the tabs

GLUE

Neck

Legs go inside this end

Legs

Legs

Capsid

Base Plate

X

|

LAB SAFETY

The Pipette Kids have forgotten their lab safety gear! Choose the clothes that will protect them from any hazards in the lab. Dress them by cutting on the bold lines and folding on the dotted lines.

|

|

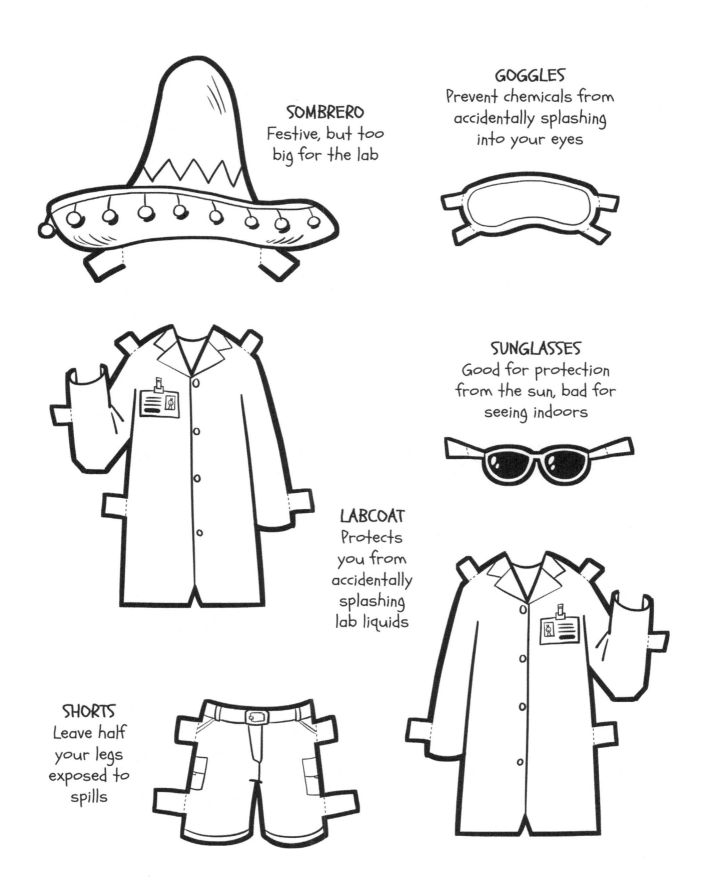

SOMBRERO
Festive, but too
big for the lab

GOGGLES
Prevent chemicals from
accidentally splashing
into your eyes

SUNGLASSES
Good for protection
from the sun, bad for
seeing indoors

LABCOAT
Protects
you from
accidentally
splashing
lab liquids

SHORTS
Leave half
your legs
exposed to
spills

|

LAB APRON
Made of thick plastic that repels liquids

PARTY HAT
For use on birthdays, at birthday parties

RUBBER GLOVES
So you don't have to touch infectious agents and toxic materials with your skin

OPEN-TOED SHOES
Sounds nice for the beach

LONG PANTS
Cover your whole legs!

CLOSED-TOED SHOES
Prevent hazardous liquids from getting all over your feet

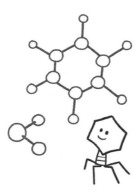

Book proceeds support science outreach educational programs for K-12 students conducted by the Graduate School of Biomedical Sciences at the University of Texas Health Science Center at San Antonio.

Funding for this book was made possible by the President's Translational and Entrepreneurial Research Fund. Special thanks to Dr. David Weiss, Dean of the Graduate School of Biomedical Sciences and Dr. Andrea Giuffrida, Vice President of Research at The University of Texas Health Science Center at San Antonio for all the support.

For more information, visit www.pipettekids.com.

Creator
Charlotte Anthony

Editor
Ramaswamy Sharma, Ph.D.

Writers
Charlotte Anthony, Armando Murillo, and Alison Doyungan Clark, Ph.D.

Layout & Design
Chase Fordtran

Illustrators
Rebecca Osborne & Angela Gao

Medical Illustrator